T0133881

Early Equipment Management (EEM)

Continuous Improvement for Projects

Early Equipment Management (EEM)

Continuous Improvement for Projects

By

Dennis McCarthy

CRC Press is an imprint of the
Taylor & Francis Group, an **informa** business
A PRODUCTIVITY PRESS BOOK

CRC Press
Taylor & Francis Group
6000 Broken Sound Parkway NW, Suite 300
Boca Raton, FL 33487-2742

Printed on acid-free paper

International Standard Book Number-13: 978-1-138-21789-8 (Hardback)
International Standard Book Number-13: 978-1-138-40016-0 (eBook)

Library of Congress Cataloging-in-Publication Data

Names: McCarthy, Dennis, author.
Title: Early equipment management (EEM) : continuous improvement for projects / Dennis McCarthy.
Description: Boca Raton : CRC Press, [2017]
Identifiers: LCCN 2016050903| ISBN 9781138217898 (hardback : alk. paper) | ISBN 9781138400160 (eBook)
Subjects: LCSH: Project management.
Classification: LCC HD69.P75 M334 2017 | DDC 658.4/04--dc23
LC record available at https://lccn.loc.gov/2016050903

Visit the Taylor & Francis Web site at
http://www.taylorandfrancis.com

and the CRC Press Web site at
http://www.crcpress.com

Printed and bound in the United States of America by
Edwards Brothers Malloy on sustainably sourced paper

Contents

Preface

As a young industrial engineer, I was lucky enough to join a company that had been acquired by new owners who were keen to grow the business. I learned lots about projects, made a few mistakes on the way, but mostly did OK. I eventually progressed to a role as project manager in Europe for a U.S. multinational, and I worked alongside many fantastic managers on Europe-wide projects to help them cope with a combination of explosive growth and economic downturns.

After a break from the world of projects, I was lucky enough to partner Peter Wilmot in his quest to bring Total Productive Maintenance (TPM) to the Western world in a way that suited Western cultures. During that time, I visited and worked with some of the best organizations in the world and was fascinated by the way they approached their business challenges. Because of my background in projects I was drawn to the way in which the best in the world approached capital investment projects and change management.

The best companies (Western and Japanese) applied a collaborative culture with robust communication between functions and levels. Their preference for teamwork and collaboration was in stark contrast to many less successful organizations where working relationships were kept at arm's length and individual agendas were more typical.

In the process I also found that some of the theoretical models developed to guide decisions about equipment were flawed and as a result led to mistakes and missed opportunities. Unfortunately, some of these models are so deeply rooted in our psyche that they are accepted as unquestionable truths even when they fail to deliver results. This means that at the early steps of a project, when judgment is needed most, the accepted wisdom can result in a poor decision.

For example, the model of quality improvement I learned as an industrial engineer was that each per cent of improvement in quality cost more than the last. The logical conclusion from this model is that there is some optimum point at which the pursuit of lower defect levels is not worth the effort. Successful organizations now accept the idea of right-first-time quality and strive for zero defects. These organizations have found that at every level of improvement, they learn more about how to improve process control and, with it, material yield. The pursuit of zero defects also holds the key to being able to make smaller batches economically, which has a major impact on cash flow items such as stock control. One of the major gains is the increase in mean time between interventions, which for some organizations has completely transformed their working practices. These are organizations that can flex production to meet peaks in demand without additional labor. As a result, their equipment has a smaller footprint, takes up less room, and costs less to buy and run.

Early equipment management (EEM) challenges a number of other accepted but false models/rules of thumb that are traditionally applied to the world of capital projects.

This book is a distillation of project work with successful organizations including 3M, BP, General Motors, Ford, GE, Ikea, Rolls Royce, Johnson Matthey, Heineken, and more recently, Princes Foods, part of Mitsubishi. I owe a debt of gratitude to many people from those organizations for the case studies and insights gained working alongside them. As part repayment of that debt I set out to provide a guidebook for others on the same journey.

For that reason, the content of the book explains *why* some of the accepted rules of thumb don't apply, as well as *how* to apply the hugely successful principles and techniques of EEM. Both of these perspectives are necessary to be able to identify weaknesses and enhance current practices to deliver the EEM goals of flawless operation from day one and ongoing low life cycle costs.

The rigor of EEM and its approach to codifying tacit knowledge is what helps to tip the balance in favor of a project's success at the start. Its range of tools and techniques for design, specification, and project management throughout the journey provide the basis for a project governance approach to increasing *value added* and *return on investment* through to day-one production and beyond.

1

Early Equipment Management: Delivering Capital Projects Faster, Cheaper, Better

Most managers and engineers have had firsthand experience of capital projects that failed to live up to expectation when introduced and which needed significant attention during routine operation. The excess costs of these troublesome assets can be huge, and not just in terms of capital costs. One study of capital projects estimated that the additional attention needed to deal with troublesome assets and subsequent loss in performance can require around three years before asset performance is sufficient to begin the planned investment payback. This is a fairly widespread problem that has been with us for some time.

Based on publicly available data, Ross Henderson (1971)* calculated the respective returns on invested capital of the Fortune 500 industrials. From the analysis, he concluded that there was "a massive failure among capital expenditure plans in North American industrial companies to provide the returns on investment which have been forecast or budgeted." Studies by Pruitt and Gitman (1987)† of Fortune 500 industrials showed that 80% of the respondents (121) admitted to having achieved lower returns than forecast, the worst results being investments in advanced technologies and new processes. This is not just a problem in larger organizations. More recent investigations in Cyprus by Lazaridis (2004)‡ focused on small- and

---

* Ross Henderson, Improving the performance of capital project planning, *Cost & Management*, Volume 45, 1971, September–October, p. 34.

† Cf. Stephen W. Pruitt and Lawrence J. Gitman, Capital budgeting forecast biases: Evidence from the Fortune 500, *Financial Management*, Volume 16, 1987, Spring, p. 47.

‡ Ioannis T. Lazaridis, Capital budgeting practices: A survey in the firms in Cyprus, *Journal of Small Business Management*, Volume 42, Issue 4, 2004, pp. 427–33.

medium-sized companies. Of the 100 studied, only a third achieved the expected return on investment.

Get it right and the gains can be significant. An oil and gas extraction company investing in a floating platform to extract oil and gas from under the Atlantic estimated that the additional output produced by achieving the early equipment management (EEM) goal of "flawless operation from day one" was enough to recoup the total capital investment costs in the first year of operation.

This chapter summarizes the main themes of the book and how the recipe for capital project success combines hard/technical and soft/collaborative skills. It also highlights the role of project governance as a vehicle for improving internal management systems and for developing the operational capabilities needed for best-in-class business performance.

1.1 WHAT GOES WRONG?

Research into the causes of underachieving returns on investment indicates that systematic front-end engineering design (FEED) processes improve capital cost, timescale, and operational performance.*

However, this is not the full picture. Although FEED weaknesses are significant, they do not account for the following frequently occurring problems:

- Conflicting views of what is needed
- Difficulties in releasing resources
- Lost opportunities to challenge and optimize design choices
- Critical decisions delayed or not taken
- Communication between project stakeholders interrupted or lost

These project governance issues frequently surface during projects, creating barriers to the delivery of results. In spite of this, they are specifically

* Gibson, G. and Hamilton, M. (1994) Analysis of pre-project planning and success variables for capital facility projects. A report to the Construction Industry Institute, University of Texas at Austin, source document 105; Gibson, G. et al. (1994) Perceptions of project representatives concerning project success and pre-project planning effort. A report to the Construction Industry Institute, the University of Texas at Austin, source document 102; Gibson, G. and Griffith, A. (1998) Team alignment during pre-project planning of capital. A report to the Construction Industry Institute, University of Texas at Austin, research report 113–12.

TABLE 1.1

Capital Project Steps

	Title	Content
1	Concept	Development of the project idea
2	High-level design	Approval of funding
3	Detailed design	Selection of vendors and detailed planning
4	Prefabrication procurement	Preparation of site and manufacture/procurement of equipment
5	Installation	Position and connect equipment
6	Commissioning	Set up and run equipment and validate process capability

excluded from the scope of project management methodologies such as PRINCE 2* (Projects in Controlled Environments, version 2).

Table 1.1 describes the EEM project steps. Figure 1.1 explains how common pitfalls at each step contribute to poor project results.

In this diagram, the lower curve illustrates the steps at which changes occur. Changes during steps 4 through 6 are in response to issues identified after the design is frozen in step 3. The upper curve shows the impact of those changes on total project costs. The shape of each curve is based on actual data captured by a machine tool manufacturer as a measure of project performance before the adoption of EEM principles and techniques.

1.1.1 Steps 1 and 2 before EEM

1.1.1.1 Change Curve

At the beginning of the project, attempts are made to obtain information to create a specification in detail.

1.1.1.2 Cost Curve

A cost estimate is made, including a contingency sum to cover unexpected costs.

* *Managing Successful Projects with PRINCE2*, Section 2.2, "Scope of PRINCE2" excludes people management, generic planning techniques, corporate management QA systems, budgetary control, and earned value analysis.

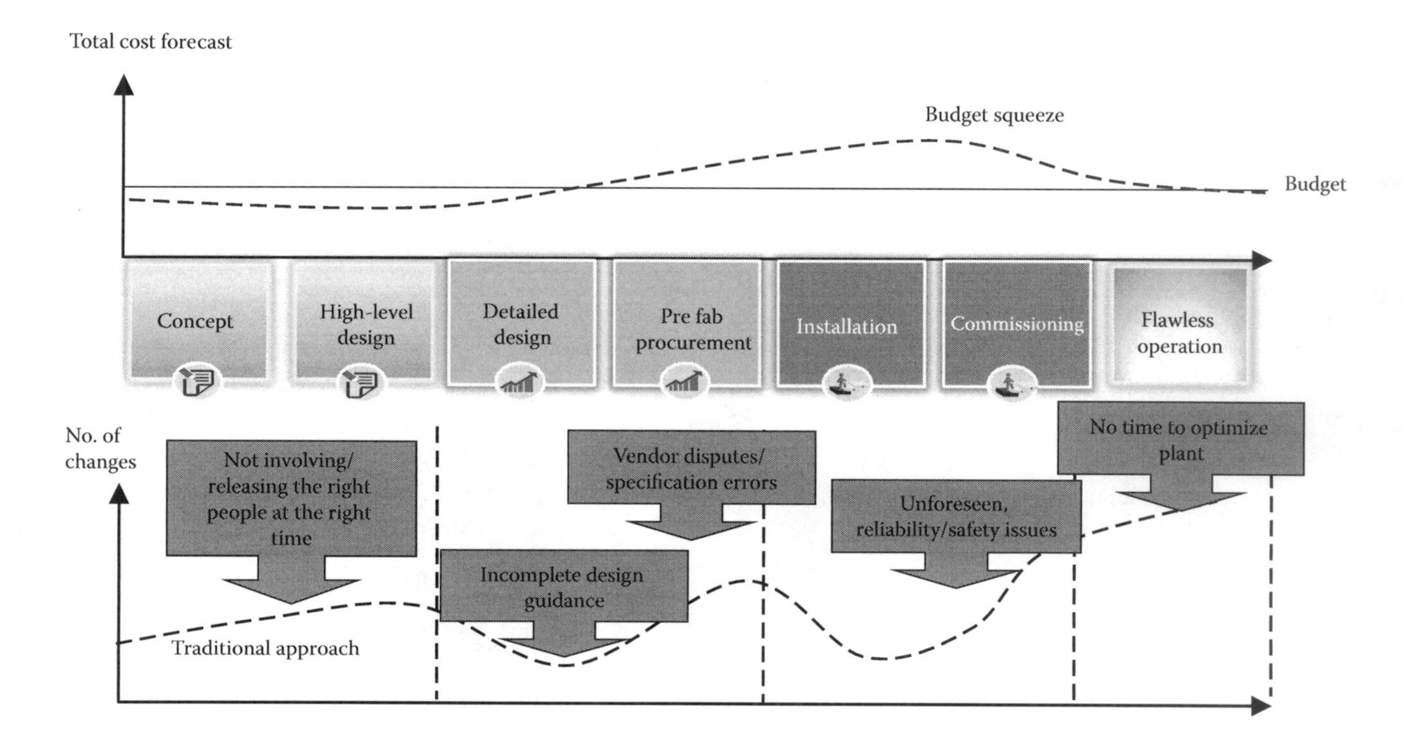

FIGURE 1.1
Traditional capital project delivery.

1.1.2 Steps 3 and 4 before EEM

1.1.2.1 Change Curve

The focus of attention during procurement is on reducing cost (vendor margins) and ring-fencing risks. Changes after the award of contracts are resisted to avoid budget creep.

1.1.2.2 Cost Curve

Unexpected (though predictable) issues arise at factory acceptance testing (FAT) prior to the dispatch of the equipment from the manufacturer, which results in additional costs and potential budget overspend.

1.1.3 Steps 5 and 6 before EEM

1.1.3.1 Change Curve

Additional modifications are needed to deal with the issues identified at FAT. This includes compromises to operational performance, which reduces the expected performance gains.

1.1.3.2 Cost Curve

Budgets are squeezed to achieve savings in *discretionary* areas such as spare parts and training.

1.1.4 What Is Really Happening

As the project passes through each step, each decision has an impact on those made at later steps. At the earlier steps, detailed information is not generally available, so some decisions are better made later. Taking detailed decisions too soon increases the risk of error. That means that the design process is iterative rather than linear. Past decisions are revisited once details become clearer. Naturally, some decisions—for example, those that define the project scope—remain fixed. Others can be taken later without risk to project performance.

For example, at the early steps of a personal project, such as a move to a new property at the concept step, choices will be made against broad criteria, such as the need to be close to work locations or family, the availability of schools for children, and so on. This choice will guide the evaluation of options. Assume that our concept scope leads us to conclude that we need a town location, the next step will involve a review of the available town-based housing stock. We need to have some idea of what can be afforded, but this will be revisited once we know more about what is available. We may also want to research the suitability of areas. In defining a town location and exploring options, we are discounting properties in the country, those away from good transport links, and so on. If we had chosen a country location, the type of housing stock would be different, and our choices would be made using different criteria.

In this example, early choices are guided by a concept of what we want to achieve at the end of the project. Defining this clearly helps to unify the decision stakeholders, so that when the best option is found, it is easier to reach a decision. This definition then guides the high-level design decisions about where to look and what we can afford. This in turn will set the frame of reference when detailing what to buy and when/how to move in.

1.1.5 Voyage of Discovery

The real world of projects does not fit the logical (but false) linear model of project management.

The detail of the final project shape is revealed a step at a time as choices are made based on experience and learning. It is a process characterized by increasing granularity of decisions at each step, a voyage of discovery (Figure 1.2).

During the early steps, when less quantifiable data is available, decisions have to be based on strategic criteria and judgments. It can take time and a number of learning cycles to develop the insight to be able to make those judgments.

1.2 EEM IN ACTION

Figure 1.3 shows how the same company approached similar projects after adopting EEM principles and techniques.

1.2.1 Steps 1 and 2 with EEM

1.2.1.1 Change Curve

An operational review at the start of the project helps to set and refine the project brief. This highlights internal knowledge gaps and the resource skill profiles needed to support the first steps toward the desired outcome.

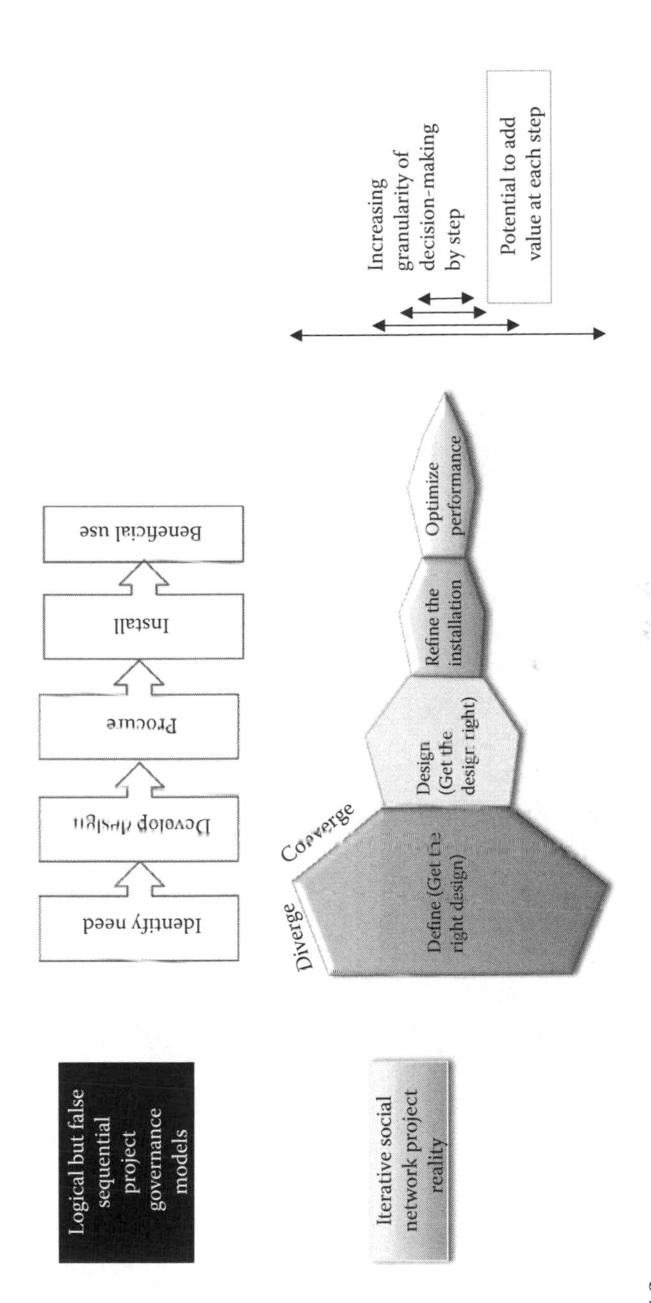

FIGURE 1.2
Iterative project reality.

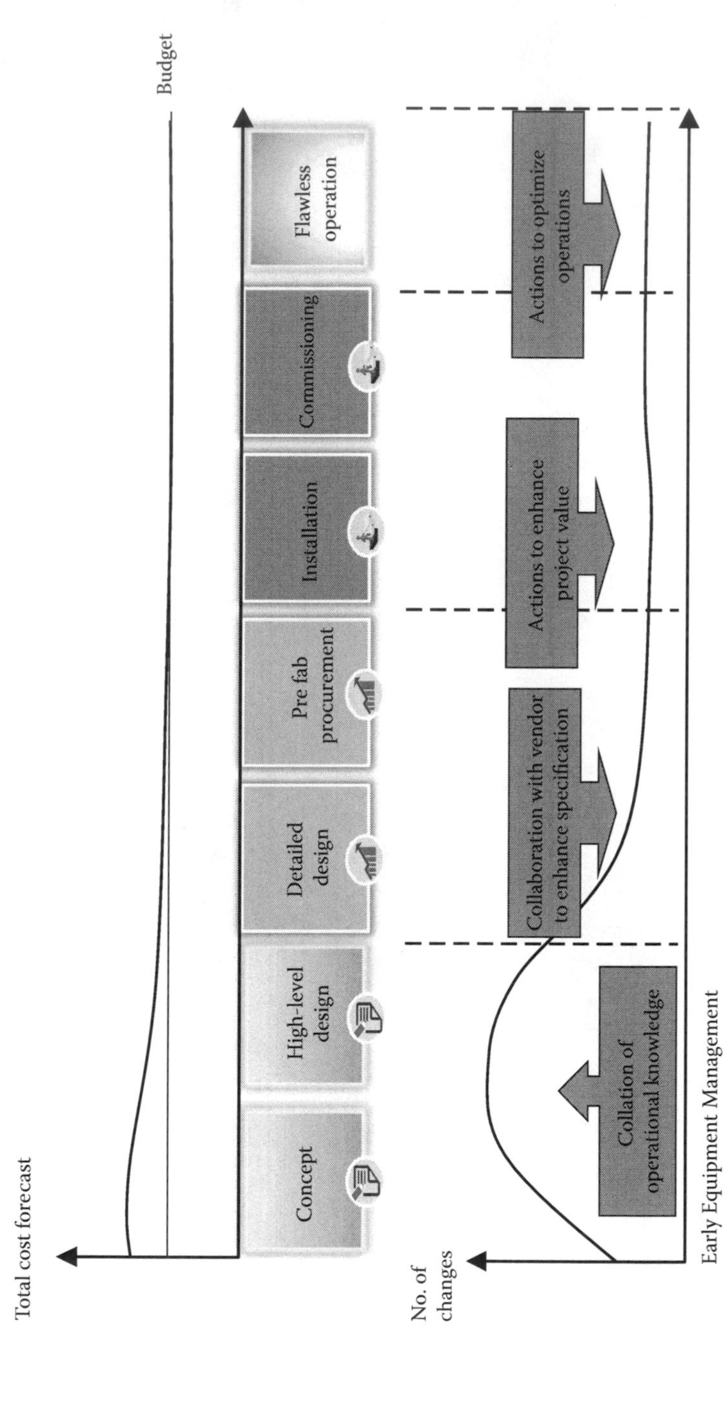

FIGURE 1.3
The EEM project delivery process.

On some occasions, this could mean collaborating with strategic customers, suppliers, and/or equipment vendors. With the right mix of skills and experience, design module reviews are used to firm up design goals, guidelines, costs, and time lines.

1.2.1.2 Cost Curve

The cost curve estimate is based on *life cycle cost* (LCC; i.e., capital costs plus ongoing operational costs). Budgets are set with the expectation that continuous improvement tools will be applied during the project to identify ways to reduce LCCs and increase project value in target areas. These goals are included as part of the funding approval process.

1.2.2 Steps 3 and 4 with EEM

1.2.2.1 Change Curve

Vendors are selected to complement internal skill sets. Contract terms include expectations of collaboration and the sharing of added value gains. Design guidelines are used to refine ideas and firm up the value-adding specification.

The design process includes detailed forward planning and agreements on *witnessed inspection/project quality plans* to ensure the glide path to day-one production.

In parallel with vendor manufacturing of the asset, work commences on operational change management and the engagement of those who will be involved in the new operation. This includes methods development, skill development cascades, and the completion of witnessed inspection testing. Lessons learned during the process are transferred to current equipment where possible. These gains are also credited to the project.

1.2.2.2 Cost Curve

Project cost tracking includes details of project added value and the results of witnessed inspection as a measure of the probability of cost achievement.

1.2.3 Steps 5 and 6 with EEM

1.2.3.1 Change Curve

Witnessed inspection tests confirm the glide path to flawless operation. Working methods are refined as part of the installation and commissioning process. As equipment is installed, the detailed workplace layout, setup routines, and operational visual aids are refined. Training cascades are used to develop the core competencies of operational personnel prior to day-one production. Skill development plans are updated to support the development of intermediate and specialist skills during the post-day-one stabilization step. The project team focuses on developing the road map to optimized operations.

1.2.3.2 Cost Curve

Project cost tracking includes details of the project's added value and the results of witnessed inspection as a measure of the probability of cost achievement.

1.2.4 The Gains

Financial gains reported as a result of using EEM principles and techniques include:

- Investment avoidance
    - A successful national commodity operator improved existing assets to achieve £500,000 investment avoidance.
    - A leading aero engine manufacturer resolved nine current equipment design weaknesses to increase capacity at a bottleneck process by 25% for less than 5% of the option to achieve that capacity through a capital investment of £17 million.
- Improved project added value
    - A waste disposal company designed equipment with a 30% reduction in minor stops and a 30% increase in mean time between failures (MTBF).
- Faster speed to market
    - A multinational innovation leader increased the capacity of a new product line by 25% and improved material yield to reduce LCC by 10%.

In addition to these financial gains, EEM benefits include:

- Improved collaboration across functional/company boundaries
- Better, slicker, simpler processes for project design, specification, and management
- The unlocking of tacit operational knowledge

- Clear investment priorities
- Increased innovation
- The proactive ownership of strategic goals

1.3 WHY WE NEED EEM

It would be easy to assume that the word *early* in the term *early equipment management* refers only to the front-end design/early project steps. On the contrary, the term refers to the principle of trapping problems as early as possible in the project process, when they are cheapest to resolve. Although it is true that the early steps of the project have a major impact on capital value, there are pitfalls to avoid and value to be added at each step of the project delivery process.

1.3.1 Avoiding Project Delivery Pitfalls

Capital project delivery involves the use of three interlinked subsystems. These are

- Design and performance management
- Specification and LCC management
- Project and risk management

Table 1.2 illustrates the nature of weaknesses or common pitfalls for each process under a traditional capital project delivery approach.

A study of these weaknesses and how to avoid their impact on project delivery is what led to the development of EEM principles and techniques. Chapter 2 explains how these three processes are combined into a single integrated project delivery game plan. The individual processes are covered in more detail in Chapters 3 through 5.

1.3.2 Making Better Decisions

The EEM project delivery process works best in organizations where the internal management processes are characterized by

- Effective working relationships
- Interaction and networking (across departments and levels)
- Clear strategic intent and control
- Collaborative learning and the sharing of ideas

TABLE 1.2

Common Project Weaknesses

	Weakness	Symptom	Why Is It Bad?
Specification weakness	(a) Investment definition	Incomplete or flawed project briefs	This contributes to late changes to specifications, unrealistic requirements for the funds/resources available, or project underdelivery.
	(b) Resource allocation	Lack of skilled resources or knowledge gaps	There is an increased probability of opportunities/issues going unrecognized or increased time and cost to deliver results.
Design weakness	(c) Design process weakness	New equipment that fails to prevent known problems or reduces performance in the first six months	This will divert resources from other parts of the business, resulting in higher operating costs and a loss of competitive capability.
	(d) Collaboration weakness	Equipment arriving on site without adequate planning	This is a symptom of poor working relationships/ collaboration. This can mean that weak ideas are untested until it is too late, resulting in unforeseen (but predictable) installation and commissioning problems.
Project management weakness	(e) Equipment complexity	Difficult to operate/ maintain assets, complex processes with residual safety and environmental weaknesses	Project management processes that don't incorporate activities to challenge design weaknesses and refine specifications overlook opportunities to achieve higher project added value and return on investment.
	(f) Project handover	Time taken debugging after the introduction of new equipment	Late debugging is costly and indicates weaknesses in earlier project stages.

These attributes impact on the effectiveness of two further EEM subsystems that create an environment in which achieve the project's full potential. These are

- Project governance
- Knowledge management

1.3.3 Improving Project Governance

Project governance is used to convert unfocused discussions about opportunities into winning ideas, practical investment propositions, and ultimately, improved sources of income (Figure 1.4).

In this environment, although data is important, the use of judgment and experience is also essential. This may seem to fly in the face of accepted wisdom, but even the complex mathematical models used to mine big data and provide insight into its hidden treasure involve the use of judgment to decide which data to use and which not to use. Deming, who is quoted as saying, "In god we trust but all others bring data," advocated the use of

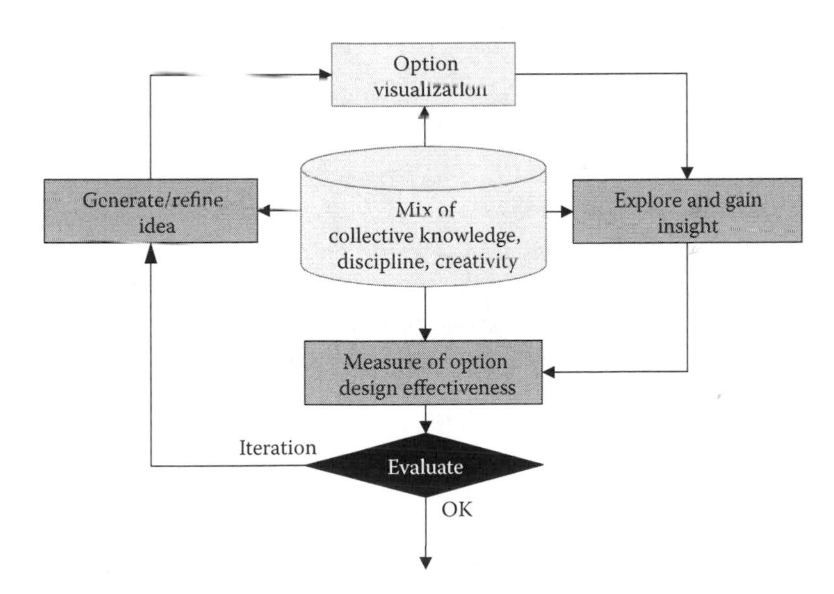

FIGURE 1.4
The social nature of projects.

judgment and learning through iterative practical application to gain the insight to make better decisions.*

Project governance provides the guidance to navigate three different types of decision landscapes on the journey from investment decision to day-one production and beyond.

These three landscape are as follows:

- *Gist-level decision landscapes*: Where decisions are based on experience, insight, and instinct. This is often characterized by language such as "That feels right" or "I am not happy about that."
  - This type of thinking is used during steps to *define* the project scope and deliverables.
- *Qualitative decision landscapes*: Where judgments are characterized by language such as *better* or *worse*. For example, *x* is quite good but not as good as *y*.
  - This type of thinking is used during steps to *design* the solution to deliver the benefits sought.
- *Quantitative decision landscapes*: Where decisions can be made on a numerical basis. For example, the output of A is four and B is three, so option A has a greater output than B.
  - This type of thinking is used to manage the delivery and *refine* the way the solution is applied to deliver the best possible value.

It is true that gist-level thinking can be subject to unhelpful biases and false memory, which if uncontrolled could contribute to project failure.

This is less of a problem in the house example, where there will typically be only one or two main decision makers whose instincts and gut feelings are reinforced with an understanding of the day-to-day detail. Discussing options together means that they can take into account multiple factors and reach a compromise on, for example, the size of the garden or transport routes because they understand the strategic implications of the choices they make.

In a typical business unit of 50–200 personnel, the insight needed to weigh up such compromises will be spread out among many people. This can be likened to completing a jigsaw puzzle when each person has only a couple of jigsaw pieces and no one has seen the picture on the outside

* Deming makes use of "plan, do, check, and act" (PDCA) action plans to test ideas and improve them.

of the box. Furthermore, because of our desire to innovate, it is a jigsaw puzzle with more than one potential outcome.

Unfortunately, without a good understanding of the day-to-day realities, it is difficult to be sure that decisions taken at the early stages do not discount viable options. To paraphrase George Santayana,* a respected twentieth-century philosopher, those who cannot remember the past are condemned to repeat it. Being able to capture and codify knowledge to guide decisions is an important enabler to secure high levels of return on investment.

1.3.4 Improving Knowledge Management

The knowledge that makes the difference between good and great designs is tacit† rather than explicit. Applying that detailed knowledge is essential, but it needs to be in the right form to support the relevant decisions landscape at each step. At the early stages, we need to maintain a helicopter's view of what is possible to understand where the best opportunities lie, what is critical, and where there are knowledge gaps so that we can direct our resources to dealing with those priorities.

At the early implementation stages, a set of six EEM design goals are used to support the evaluation of top-level features and options. An example of the six EEM design goals and an assessment scale is set out in Appendix A. This is covered in more detail in Chapter 3, "Design and Performance Management." As projects progress to higher levels and more detailed design steps, the design goals are complemented by codified tacit knowledge in the form of design guidelines and standards. These are also discussed in more detail in Chapter 3. These design guidelines provide a way of calibrating the use of subjective terms such as *good, better, best.*

To overcome the inconsistencies of subjective accuracy,‡ design guidelines incorporate the perspective of commercial, operational, and technological functions as three separate but complementary design lenses. These three perspectives are also used to define the project team structure.

* George Santayana (1863–1952). Taught philosophy at Harvard and King's College Cambridge. Famous for aphorisms, such as "Fanaticism is redoubling your efforts after you have forgotten your aim."

† *Tacit knowledge* explains the difference between the insight and experience of a skilled and unskilled person. It is difficult to describe and transfer. By its very nature, it is continually being added to as people learn more and advances are made in technology.

‡ Daniel Kahneman and Gary Klein, Conditions for intuitive expertise: A failure to disagree, *American Psychology*, Volume 64, Issue 6, pp. 515–26.

Within the EEM core team, each function has an equal opportunity to define design goals, guidelines, and standards. They also have an obligation to challenge ideas that do not meet those design goals and guidelines (see Chapter 2, "EEM Road Map").

1.3.5 Delivering Better Management Processes

To recap, the EEM process is designed to achieve the following goals:

- Flawless operation from production day one
- Lowest operational LCC

The successful application of EEM principles and techniques also improves capabilities in project governance and knowledge management.

Chapter 6, "Project Governance," includes the results of research showing how these capabilities contribute to the success of best-in-class organizations. Their ability to deliver projects well is both a litmus test of good business governance and a clear source of competitive advantage.

An EEM implementation case study, included in Chapter 7, "Implementing EEM," provides a road map to replicate the success of the project. It also explains how to lock in the gains by raising standards, achieving the project goals in less time and with fewer resources.

The benefits of extending EEM principles and techniques to the development of new products and services are explained in Chapter 8, "Early Product Management."

The outcome is increased speed to market and accelerated rates of growth, profitability, and return on investment.

This is an approach that can also be applied to a wide range of project types from capital equipment to software implementation projects.

1.4 BOOK STRUCTURE

1.4.1 Overview

The book is designed as a training and reference aid, rather than something to be read cover to cover. Each chapter is designed as a stand-alone topic with cross-references to other chapters where relevant.

Chapter 2 explains the basic EEM project delivery steps. Chapters 3 through 6 explain the underpinning systems and processes. Each chapter also cross-references the relevant EEM road map's project governance phases of *define*, *design*, *refine*, and *improve*.

Chapter 7 uses a case study to explain the implementation methodology.

Chapter 8 explains how the principles and techniques of EEM have been extended to include *early product management* (EPM), recognizing the major impact that product and service development has on equipment specification and project delivery.

1.4.2 Chapter Summaries

The following are the topics covered and key learning points for each chapter.

Chapter 2 sets out the EEM project road map. This provides an overview of

- How the EEM road map integrates project delivery, project governance, and knowledge management
- The project governance learning landscapes
- The EEM organization and a workstream RACI (responsible, accountable, consulted, informed) chart
- The responsibilities of management
- Setting out the project learning agenda
- EEM accountabilities for each step
- Signposts to other chapters covering topics in more detail

Key learning points include the following:

- Successful capital project delivery processes use structured decision phases that add detail at each step.
- Each phase covers a different decision landscape with different project governance challenges.
- EEM tools support these decision cascades across the complete concept to project the delivery process.

Chapter 3 provides an explanation of *EEM design and performance management*, the first of the five EEM subsystems, covering

- How developing good design depends on creating the conditions for innovation and processes to guide systematic design development
- How both innovation and systematic design development benefit from an understanding of
  - Current shop floor reality
  - Translating tacit into explicit knowledge
- How to manage parallel working and integrate the workflows of multiple stakeholders
- Vendor skill sets and how to get the best from working with them to tease out design weaknesses and prevent problems
- The design of activities to develop in-house capabilities and manage project-related operational changes

Key learning points include the following:

- Design is a complex *wicked problem* that benefits from a stepwise approach that considers multiple options as a way of developing a preferred one.
- Be clear about what you want and manage the design requirements of all stakeholders against a common outcome and workstream milestones.
  - Develop design guidelines to support EEM goals.
  - Create technical standards carefully so that they aid standardization and guide detailed equipment manufacture without creating a unique design proposition that excludes the use of vendor experience and innovation.
  - Break the design brief into modules to create manageable work packets and to structure the planning, organization, and control of the design and delivery process.
  - Carry out an operational review for each module to highlight potential areas of weakness and opportunity.
  - Use this to define the performance drivers for each module.
  - Add detail to and refine the design of each module, including operational ways of working through practical activities at each project step.

Extend the design process to include performance delivery, confirm design assumptions, and enhance internal design capabilities.

Chapter 4 provides an insight into *specification and LCC management*, the second of the five EEM subsystems, and covers

- Making choices, refining designs, defining equipment functions and design modules
- Documenting the preferred option to enhance project value
- Working with vendors

Key learning points include the following:

- Build a collaborative relationship based on shared goals.
- Add detail to the specification at each step so that it evolves into an operations manual.
- Develop checklists based on the issues raised to support the quality plan/witnessed inspection.
- Focus on reducing the LCC to get the best possible value from the investment in both money and brain power.
- Understand equipment losses and countermeasures and their impact on LCC.
- Treat each project as a voyage of discovery, using the LCC model as the hidden-value treasure map.

Chapter 5 explains *project and risk management*, the third of the five EEM subsystems, and covers

- Project plan development so that plans are realistic and achievable. Also covered are the use of different formats, including the scheduling of detailed work packets during installation and commissioning.
- Managing project teams, including how to facilitate/coach teams and create a positive working environment so that they can achieve more.
- Managing risk, including the identification of hazards, the timing of risk assessments, and the use of witnessed inspections to manage the glide path to flawless operation.
- A project leader's guide to EEM project delivery.
- Links with other chapters, including specification management and project governance.

Key learning points include the following:

- Use milestones supported by detailed activity plans to guide projects through the EEM process.
- Set exit criteria for each milestone stage gate.
- Assess risk and use witnessed inspection to confirm the achievement of the project quality plan.
- Project managers need to become project leaders to get the best out of the core and stage gate teams.
- The glide path to flawless operation should be managed as closely as an aircraft coming into land or a spacecraft landing on the moon.

Chapter 6 explains *project governance* and *best-practice design books*, the fourth and fifth EEM subsystems, and covers

- How EEM supports improved strategic control
- Support for the leadership challenge of improving capabilities by providing the opportunity to develop those taking part in projects and those who backfill their positions
- The benefits of closer working relationships with vendors
- Progress measured through changes in outlook
- Incorporate EEM within the organization's continuous improvement process to improve current asset performance and raise standards to support higher levels of technical stability and operational resilience

Key learning points include the following:

- Develop a site *EEM master plan* defining commercial, operational, and technological challenges.
- Align these with site-strategic goals, guiding capital investment priorities and setting the context for the scope of capital projects.
- Develop design guidelines, technical standards, and checklists to codify experience, guiding design decisions and project management delivery.
- Create LCC models to identify cost drivers and support option evaluation.
- Manage the path to flawless operation.
    - Installation and commissioning
    - Developing operational capability

- Use the stabilization step to define the road map to optimum conditions. Make this the launchpad to better-than-new performance.

Chapter 7 explains the EEM implementation process and covers

- EEM implementation steps
- Implementation milestones
- An implementation case study
    - EEM strengths and weaknesses diagnostic
    - Management awareness
    - Mobilization
    - Pilot project
    - Policy development
    - Roll-out cascade
    - EEM learning plans and competency assessment
- Speeding up time to market

Key learning points include the following:

- Measure organizational strengths and weaknesses using EEM audit criteria.
- Use practical pilot projects to learn how to apply EEM principles and techniques.
- Create an EEM learning plan for key stakeholders and project team members.
- Support their efforts to integrate the way projects are run into the business routine.
- The success of EEM depends on knowledge sharing and collaborative behaviors.

Chapter 8 explains EPM and covers

- What EPM is and why we need it
- EPM steps
    - *Shell*: innovation generation
    - *Shape*: idea development
    - *Scope*: product/service design
- Design, specification, and project management approaches to support each step

- Links with EEM, including EPM project governance issues
- Implementing EPM

Key learning points include the following:

- The product defines the capital, which defines the operation.
- Integrate shape and EEM concept activities to improve the design for manufacture.
- Collaborate with strategic partners to refine good ideas into winning products and services.
- Balance commercial and technological progression to allocate scarce resources to product and service offerings with the best chance of success.

2

The EEM Road Map

The EEM road map integrates five underpinning EEM subsystems (Table 2.1) into a seamless integrated project delivery process, one that is capable of achieving flawless operation from day one and the lowest life cycle cost (LCC). To use a sporting analogy, the road map sets out the purpose and rules of the game. The subsystems define the competencies and tactics that make the difference between winning and losing. Project and risk management sets out the team roles and formation. Design and performance management provides the game-winning attack. Specification and LCC management are the impenetrable defense. Project governance and knowledge management are the backroom boys supporting the efforts of the team to score more points. These are described in Table 2.1.

As with a top-class sports team, there are specialists. The EEM team consists of players with commercial (customer facing), operational (order fulfillment), and technology-related expertise. Just like a top-class sports team, all players need to be on the field of play throughout the match and able to play in attack or defense as needed. This chapter sets out the EEM game plan, how it works, and how to apply it to ratchet up return on capital investment.

Design, specification, and project management subsystems are covered in more detail in Chapters 3 through 5. Project governance and knowledge management processes are dealt with in Chapter 6.

2.1 EEM ROAD MAP OVERVIEW

The EEM road map is designed to trap latent weaknesses at each step, starting with the concept development and ending with the handover to operations after the achievement of stable operation.

TABLE 2.1

Five Subsystems of the EEM Road Map

EEM Subsystem	Purpose
1. Design and performance management	The systematic identification, analysis, and selection of design features to enhance project value, reduce project risks, and lower total LCCs.
2. Specification and LCC management	The timely capture and formatting of key information at each step as a resource to provide insight, aid communication, and support the delivery of project goals
3. Project and risk management	The planning, organization, and control of project resources to trap and resolve latent problems and risks early, and to deliver project goals using the expertise of the complete internal and external team
4. Project governance	The quality assurance of each EEM step to prevent problems from being transferred to the next, and to support the achievement of a) flawless operation from day one and b) the lowest possible LCCs
5. Best-practice design books	The collation and codification of knowledge to guide the delivery of low LCC designs

TABLE 2.2

EEM Project Governance Phases and Project Steps

Title		Goal
Define		**To get the right design**
1	Concept	Development of the project idea
2	HLD	Approval of funding
Design		**To get the design right**
3	Detailed design	Selection of vendors and detailed planning
4	Prefabrication procurement	Preparation of site and manufacture/procurement of equipment
Refine		**To get the design gains**
5	Installation	Positioning and connecting equipment
6	Commissioning	Setting up and running equipment and validating process capability

The EEM road map steps in Table 2.2 are similar to the steps set out in international standards such as ISO 9001.

Each pair of steps are grouped under EEM project governance phase headings. Each phase covers a different part of the project governance journey. The odd-numbered steps are the midpoint or half-life signposts on that journey. The project governance phases are discussed in more detail in the following sections.

2.1.1 Define: Getting the Right Design

The aim of this phase is to define what is needed and how much can be spent to deliver it. The level of detail should be sufficient to support outline discussions about what is possible and what options are available. That should include a good level of awareness of the current operational reality to understand how the available offerings will impact on it.

This phase is characterized by questions such as, What do we want, what do we need, what are the options, what can be achieved, and what can we afford? This is similar to the house-hunting quest from Chapter 1.

Avoid the common pitfall of having too narrow a frame of reference. Use innovation-friendly processes to consider a wide range of options before homing in on a short list of viable options. Use design goals and guidelines (see Chapter 3, "Design and Performance Management") to select and refine a preferred option. This can be a relatively informal and quick process, but give people the time and space to reflect on and refine decisions before progressing to the high-level design (HLD) stage gate review.

Table 2.3 sets out the EEM tools that may be used during this phase.

TABLE 2.3

Define Phase EEM Tools

Tool	Application
EEM diagnostic	Assessing strengths, weaknesses, and potential gains from applying EEM principles and techniques
EEM vision and project scope definition	Action mapping
EEM milestone plan	Sets out future tasks, stage gates, and resource needs
Voice of the customer	Defining the voice of the customer; monitoring and reporting on changes in market requirements
DILO (day in the life of) review	Task definition, job design, operations organization, definition of production and maintenance methods, and asset conditions
PP data (aka maintenance prevention data)	Understanding current asset-hidden losses, common problems, and countermeasures; defects, design weaknesses, and improvement potential
Criticality analysis	Understanding current design weaknesses; defining design modules and priorities for attention
Module review	Assessing and refining the preferred concept, identifying risks/PP tactics, and developing commissioning test protocols
EEM design goals, guidelines and standards	Used to codify tacit knowledge and lessons learned; supports option evaluation, vendor selection, detailed design, risk assessment, and PP processes
Objective testing	Used to short-list viable options from a long list, compare option attributes, and test those attributes to identify and refine a preferred option

The outputs from this phase typically include

- Approved funding to justify delivery of a defined business benefit
- A statement of performance targets and deliverables
- A forward outline plan and resources for the next step
- A short list of potential vendors

2.1.2 Design: Getting the Design Right

The aims of the design phase are to select the right partners, agree contractual terms, and collaborate with them to develop realistic and achievable action plans. This phase is characterized by questions such as, Who can best meet our needs, what are the risks, what do we need to learn, how can we work with them, and how will we make this happen?

Avoid the common pitfall of assuming that the selected vendor understands your business. Recognize that the vendor sales team members are unlikely to be involved in the details of the vendor design, manufacturing, installation, or commissioning processes. Take the time to induct the vendor team and make sure that the vendor can demonstrate a clear understanding of your design goals and guidelines. Agree a quality plan with definitive *witnessed inspections* to assure progress and manage risks. Add to this internal operational change management.

In addition to the tools included in Table 2.3, the following tools and techniques may be applied during this phase (Table 2.4).

TABLE 2.4

Design Phase EEM Tools

Tool	Application
Operations provisional best-practice routine development	Developing draft ways of working as input to skill development and training cascades; includes transition activities/plans and outputs of risk assessments
Visual management	Actions to make normal condition, setup, and asset care parameters visible at a glance; includes actions to minimize the need for adjustment/intervention
Detailed project plan	Integration of all tasks and related projects on a single timeline
Installation/commissioning work packet design	Designing work packets to support visual management and critical path analysis
Operating manuals, technical handbooks, and learning process development	Coordinating documentation/refining operating procedures, skill provision routes, recruitment/training, confirmation of competence, and standardized operating procedures

The outputs from this phase typically include

- Readiness to install
- Witnessed inspection testing, covering installation and commissioning
- A skill development cascade
- A communications plan

2.1.3 Refine: Getting the Design Gains

The aim of this phase is to manage the transition from old to new and to refine the way the organization works to deliver maximum gains. Operationally, this can be a highly disruptive phase characterized by questions such as, Who is doing what, where do we start, how are we doing, and are we nearly there yet?

Avoid the common pitfall of relying on vendor training alone to deliver internal competencies. Manage the internal training cascade so that internal resources are able to cascade training to achieve demonstrable core competencies* prior to day-one operation. Refine workplace layouts as part of the installation so that the commissioning steps are carried out in conditions that are as near to operational readiness as possible.

Midway through installation, confirm the glide path to flawless operation and monitor progress closely. Be prepared to stop, take stock, and if necessary delay the start day or increase resources to get back on track if progress is not as expected.

In addition, the following tools and techniques may be applied during this phase (Table 2.5).

TABLE 2.5
Refine Phase EEM Tools

Tool	Application
Best-practice routine training	Pre-day-one production to support the delivery of flawless operation
Debugging reports	Stoppage analysis; project management analysis reports
Test reports	Testing results against defined standards and feedback to design best-practice books
Problem-solving sheets and situation reports	Capturing problem-solving activities and results

* Split skill levels into core, intermediate, and specialist skills. Aim to achieve core competencies before day-one production and intermediate/specialist skills during the stabilize step.

Outputs from this phase typically include

- The achievement of flawless operation
- Confirmation of core competencies
- A development cascade of intermediate and specialist skills

2.1.4 Improve

Following the refine phase, the priorities are to achieve stable running and manage the transition from project to operational status. This seventh step also includes the identification of a route to optimum conditions. That means that site acceptance testing (SAT) becomes a launchpad for the continuous journey toward optimized operations.

This handover of the improvement baton from EEM to the operational improvement process provides the lever to lock in the gains. Without this focused improvement driver, the condition of the new asset may be allowed to deteriorate. The result will be stable operation for a period of time before the process becomes unreliable and subject to frequent stoppages (Figure 2.1).

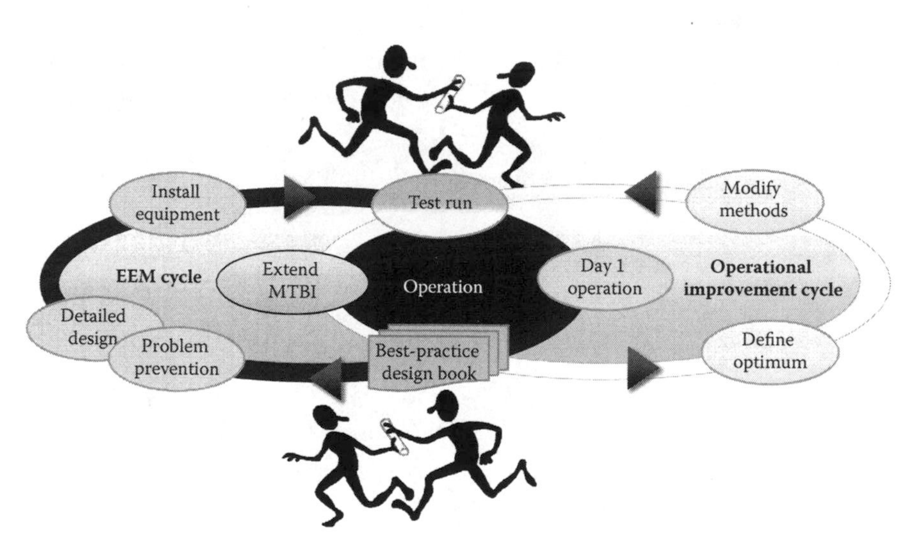

FIGURE 2.1
Handing over the improvement baton.

2.2 EEM PROJECT WORKSTREAMS RACI

Each step of the EEM road map is organized into five workstreams:

1. Mobilization, including planning the plan
2. Design decisions to be made at that step
3. Updating the specification to reflect the decisions made
4. Planning and resource definition for the next EEM step
5. The stage gate/readiness review process

These workstreams are used to explain the EEM project roles as set out in Figure 2.2.

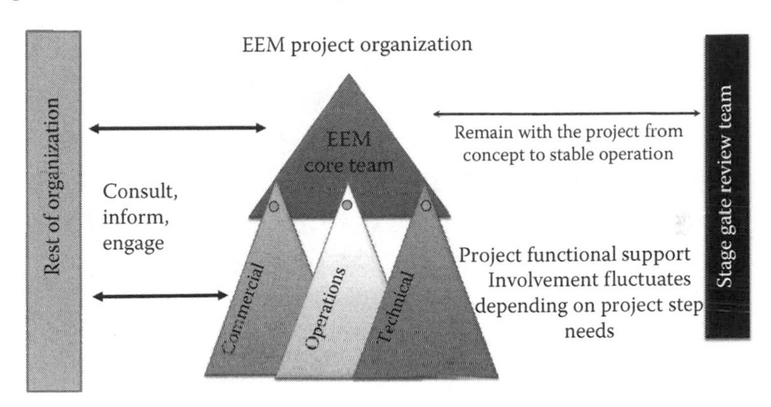

FIGURE 2.2
EEM project organization.

Table 2.6 contains a RACI (*responsible, accountable, consulted, informed*) chart explaining each EEM role against the preceding workstreams.

TABLE 2.6
EEM Workstream RACI Chart

	Responsible	Accountable	Consulted	Informed
1	Site management	Project steering	Project leader	EEM core team
2	Project leader	EEM core team	Project leader	Project steering
3	Project steering	Project leader	EEM core team	Site management
4	Project steering	Project leader	EEM core team	Site management
5	Site management	Project steering	Project leader	EEM core team

The following definition uses the original RACI concept, where "R" in the left-hand column is the most senior role.*

- *Responsible*: Sets policy, defines project briefs, and ensures understanding
- *Accountable*: Carries out actions
- *Consulted*: Available to advise
- *Informed*: Kept advised

2.2.1 RACI: Responsible

The *responsible* role supports the project governance process and provides a link between the management of the current operation and the project process.

The scope of the EEM benefits and therefore the scope of those in the responsible column include

- Development of in-house capability to deliver
  - Flawless operation from day one
  - Low operational LCCs
  - Increased return on investment
- The capture of tacit knowledge to support
  - Cross-project learning
  - Clarity of investment priorities
  - Project ownership
  - Innovation
- Improvement of project management processes to
  - Speed up project delivery
  - Improve collaboration across functional/company boundaries
  - Improve design, specification, and project management systems and processes

During the project, a key part of the responsible role is the transfer of project lessons learned to parts of the existing operation where the learning can add value.

* In some organizations the definitions of *responsible* and *accountable* are reversed. Under that definition, to retain the logic of the most senior role in the left hand column, the acronym becomes ARCI. Use whatever format is currently applied in your organization.

For example, during an EEM *day in the life of* (DILO) review for a new line, the EEM team identified a weakness in *cleaning in place* (CIP). The end-of-run procedure could on occasion result in the dilution of CIP fluids and a failure to meet lab-testing requirements at the start of the next run. A way of preventing the problem was identified for the new operation. The site management team championed the use of improved methods on existing production lines to produce a step change in production plan adherence levels.

Although capturing lessons learned is an obvious thing to do, it is a rarely completed activity. To avoid this loss of project value, carry out the following:

1. Create an *equipment management master plan* to establish a clear vision of the link between business drivers and improvement priorities (see Chapter 4, "Specification and LCC Management").
2. Identify where weaknesses in systems and working methods present barriers to performance and use them as triggers for improvement within every project.
3. Capture lessons learned as case study stories written in an easy-to-digest style to engage new project members and encourage the sharing of good ideas.
4. Train subject matter experts and new project members in the improved processes and reinforce their application during stage gate planning and stage gate review activities.

2.3 RACI ACCOUNTABLE COLUMN: DEFINE PHASE

The following is an explanation of the roles in the RACI *accountable* column covering those who carry out the work during EEM step 1 (concept development) and step 2 (HLD) (Figure 2.3).

2.3.1 Concept Development

Aim: To develop a preferred concept

1. Mobilize.
   - Clarify project scope: Define the team RACI and *management control reporting system* (MCRS).
   - Obtain the release and organize a mobilization meeting for those involved.

Tasks	Notes	Stage gate team	EEM core team	Project leader
1.1 Mobilize concept.	Clarify brief, identify people to be involved, and hold first team meeting.	◎		○
1.2 Select preferred concept.	Objective testing and selection of preferred option.	○	◎	○
1.3 Create concept specification.	DILO and design assessment.		○	◎
1.4 Develop project plan.	Action map development.		○	◎
1.5 Concept stage gate review.	Audit 1 and stage gate review.	◎	○	○
2.1 Mobilize HLD.	Prep for objective testing of HLD options.	◎		
2.2 Select concept delivery approach.	Objective testing HLD.	○	◎	○
2.3 Define HLD specification.	HLD module/value engineering review.		○	◎
2.4 Develop project plan step 2.	Action mapping for detailed design step.		○	◎
2.5 HLD stage gate review.	Audit 2 and stage gate review.	◎	○	○

FIGURE 2.3
EEM road map *define* phase quality plan.

2. Select the preferred concept.
   - Confirm project brief: Review *problem prevention* (PP) data to confirm the understanding of current operation and effectiveness-hidden losses. Define the future project vision, set targets, and develop tactics to deliver the desired outcome.
   - Generate options: Develop a long list of options for review and rationalize into a short list for detailed evaluation.
   - Evaluate the option short list; refine the preferred option.
   - Create an outline list of equipment; define civil engineering, utilities, and interfaces with current systems and processes.
3. Create the concept specification.
   - Formalize the concept specification, including the LCC model and benefits to support the development of justification.
   - Define the list of actions needed to achieve flawless operation; include a functional team assessment of red/amber flag risks.
4. Develop the project plan.
   - Develop a milestone and resource plan for each EEM step and align with the timings of other projects and events, such as planned production outages.
   - Develop the concept business case.
5. Confirm readiness.
   - Stage gate review: Confirm that the preferred option meets the design objectives and EEM audit criteria.
   - Confirm the forward plan for the next steps and resource allocation.

2.3.2 High-Level Design

Aim: To approve funding and the basis for vendor selection

1. Mobilize HLD.
   - Mobilize the team: Define the team RACI and MCRS, obtain the release, and organize a mobilization meeting.
   - Confirm the concept assumptions and list of requirements. Add detail to the preferred option and specify the relevant standards to be applied; explore options for delivering the concept.
2. Select the concept delivery approach.
   - Evaluate options and select the preferred concept delivery approach.

- Collate a list of requirements, refine the layout, and explore the impact on current operations.
- Define design modules and targets.
  - Produce an information pack for each design module, including layouts, changes in workflow, working methods, and asset footprints.
  - Produce a specification, including details of applicable design goals, guidelines, and standards in a format to support vendor evaluation.
  - Confirm risk mitigation tactics from the risk assessment, including the *factory acceptance test* (FAT), *site acceptance test* (SAT), protocols to support the project quality plan, and the Construction Design and Management (CDM) Regulations.

3. Define the HLD specification.
   - Carry out an HLD risk assessment.
   - Obtain budget estimates and create a budget for the project. Update the LCC model.
4. Develop the project plan.
   - Update the milestone quality plan, project organization, timing, and critical path.
   - Develop the activity schedule, task list, and timetable.
   - Align with the timings of other events during the proposed project timescale.
   - Define design module detailed design workshop format.
   - Develop the funding application.
5. Confirm readiness.
   - Stage gate review: Confirm that the preferred option meets the design objectives and EEM audit criteria.
   - Confirm the forward plan for the next steps and resource allocation.
   - Begin the funding approval process.

2.4 RACI ACCOUNTABLE COLUMN: DESIGN PHASE

The following is an explanation of the roles in the RACI *accountable* column covering those who carry out the work during EEM step 3 (detailed design) and step 4 (prefab procurement) (Figure 2.4).

Tasks	Notes	Stage gate team	EEM core team	Project leader
3.1 Procure.	Vendor selection and approval.	◎		○
3.2 Detailed design.	Detailed design workshops.	○	◎	○
3.3 Freeze specification step 3.	Complete risk assessment 3 and finalize specification after workshop 2.		○	◎
3.4 Detailed activity planning step 3.	Action map for installation and commissioning.		○	◎
3.5 Detailed design stage gate review.	Audit 3 and stage gate review.	◎	○	○
4.1 Mobilize PFP.	Mobilize resources and deploy/refine plans to deliver detailed design.	◎		
4.2 Operations change management step 4.	Develop operational organization, methods, and implementation.	○	◎	○
4.3 Organize installation and commissioning step 4.	Firm up installation plans and organization.		○	◎
4.4 Equipment manufacture QA step 4.	Design and implement witnessed inspection during manufacturing and pre installation stages.		○	◎
4.5 Installation readiness review/stage gate.	Audit 4 and stage gate review.	◎	○	○

FIGURE 2.4
EEM road map *design* phase quality plan.

2.4.1 Detailed Design

Aim: To develop a detailed specification and project delivery program

1. Procure.
    - Mobilize the team: Define the team RACI and MCRS, obtain the personnel release, and organize a mobilization meeting.
    - Vendor selection:
        - Following funding approval, circulate the *request for quotation* (RFQ).
        - Review vendor bids, visit reference sites, and select a preferred vendor.
        - Agree contractual terms; appoint vendor(s).
        - Place orders.
2. Detailed equipment design.
    - Induct the vendor into the EEM process, agree the project vision and route to flawless operation, and establish a foundation for collaboration.
    - Communicate change control and document management protocol version control; project portal author, editor, and access rights.
    - Detailed design workshops: Use a workshop to plan an in-depth review of detailed design modules to tease out latent design weaknesses and opportunities to add value to the project.
        - Include
            - Installation planning at this stage
            - Training needs and the training/skill development process
            - Risk assessment (detailed design sign-off)
3. Freeze the specification.
    - Formalize the specification and LCC forecast.
    - Optimize the specification deliverables; develop detailed budgets.
    - Authorize the engineering change management protocols.
4. Detailed activity planning.
    - Develop design module delivery plans.
        - Refine the milestone plans; develop activity schedules, task lists, and project timing plans.
    - Confirm business justification and project quality plan criteria, including witnessed inspection tests, checklists, and schedules

5. Confirm readiness.
    - Stage gate review: Sign off plans and quality assurance steps.
    - Confirm the forward plan for the next steps and resource allocation.

2.4.2 Prefab Procurement

Aim: To manufacture equipment and prepare the site for installation

1. Mobilize.
    - Induct additional team members to support the development of operational design and change management.
2. Operations change management.
    - Develop operational methods.
        - Develop a task master list; identify skill gaps; identify human error risks and the PP approach.
        - Build on the best-practice operation and asset care used in the detailed design workshops.
    - Update the workplace organization design, tooling, cleaning, and maintenance aids.
    - Identify opportunities to increase project value.
3. Prepare for the installation and commissioning steps.
    - Review the vendor installation plans.
        - Assess against the installation standards and procedures.
        - Refine/develop the project safety plan and information packs.
    - Identify the installation site facilities management, commissioning, and handover protocols.
    - Carry out an installation risk assessment and mitigation plan.
        - List tasks and method statements and identify risks and mitigation tactics.
        - Create health and safety plans; assign work area coordinators and toolbox meeting schedules.
4. Deliver the project quality plan.
    - Detailed activity planning and work packet development.
    - Finalize detailed installation and commissioning plans with the vendor/contractors.
    - Track the progress of the witnessed inspection/quality assurance plan, including management of the construction phase and FAT.

5. Confirm readiness.
    - Stage gate review: Sign off plans and quality assurance steps.
    - Confirm the forward plan for the next steps and resource allocation.

2.5 RACI ACCOUNTABLE COLUMN: REFINE PHASE

The following is an explanation of the roles in the RACI *accountable* column covering those who carry out the work during EEM step 5 (installation) and step 6 (commissioning) (Figure 2.5).

2.5.1 Step 5: Installation

Aim: To install equipment in the plant

1. Mobilize installation.
    - Define the team RACI and MCRS, obtain the release of resources and organize a mobilization meeting.
    - Induct new team members: Include a safety briefing and communication of how to deal with issues arising.
2. Install.
    - Manage the installation process and compliance to temporary health and safety arrangements.
    - Control the installation process. Include communication, site clearance, and witnessed inspection to confirm the installation quality.
3. Precommission.
    - Confirm the glide path to delivering flawless operation from day one.
    - Mobilize the commissioning team: Manage the transfer from installation to commissioning personnel.
4. Installation coordination.
    - Coordinate the installation and precommissioning communications plan.
    - Installation issue capture, review, and action process.
5. Confirm readiness.
    - Stage gate review: Sign off plans to deliver flawless operation.
    - Confirm the forward plan for the next steps and resource allocation.

Tasks	Notes	Stage gate team	EEM core team	Project leader
5.1 Mobilize installation.	Induct new project organization and team members to engage with and support team project goals.	◎		○
5.2 Install.	Mobilize the installation plan and address risks to on-time delivery.	○	◎	○
5.3 Precommissioning.	Handover from installation to commissioning teams.		○	◎
5.4 Installation coordination.	Manage installation schedule to account for delays and early completions.		○	◎
5.5 Confirm commissioning readiness step 5.	Audit 5 and flawless operation stage gate.	◎	○	○
6.1 Mobilize commissioning.	Induct new team members with their role and commissioning protocols.	◎		
6.2 Commission.	Complete the commissioning plan and address risks to on-time delivery.	○	◎	○
6.3 Flawless operation delivery.	Manage glide path to flawless operation.		○	◎
6.4 Commissioning coordination.	Manage commissioning schedule to account for delays and early completions.		○	◎
6.5 Confirm flawless operation day-one operation.	Audit 6 and stage gate.	◎	○	○

FIGURE 2.5
EEM road map *refine* phase quality plan.

2.5.2 Step 6: Commissioning

Aim: To deliver flawless operation on production day one

1. Mobilize commissioning.
    - Induct new team members.
2. Commission.
    - Coordinate commissioning and test process/witnessed inspection to confirm commissioning quality.
3. Flawless operation delivery.
    - Confirm the transfer and acceptance protocol has been achieved.
4. Commissioning coordination.
    - Detect defects and list outstanding items.
    - Develop day-one readiness communications pack.
5. Confirm flawless operation.
    - Stage gate review: Confirm flawless operation has been achieved.
    - Confirm the forward plan for the next steps and resource allocation.

2.6 RACI ACCOUNTABLE COLUMN: IMPROVE PHASE

The following is an explanation of the roles in the RACI *accountable* column covering those who carry out the work during step 7 (stabilize) and step 8 (optimize) (Figure 2.6).

2.6.1 Step 7: Stabilize

Aim: To establish normal conditions without specialist support and define the road map to optimum conditions

1. Mobilize the team.
    - Induct new team members.
    - Establish the route map to handover from specialist to routine operational personnel.
    - Clarify intermediate and specialist skill development routes.
2. Stabilize “normal” conditions.
    - Achieve the operation stabilization goals.
    - Stabilize performance at the agreed level with internal personnel and develop optimization tactics.
    - Update training material to capture and transfer the lessons learned.

Tasks	Notes	Stage gate team	EEM core team	Project leader
7.1 Mobilize stabilization.	Induct operations team.	◎		○
7.2 Stabilize nonnal conditions.	Manage schedule to stabilize routine operations under the operations team.	○	◎	○
7.3 Achieve technical stability.	Finalize documentation, knowledge transfer, and plans to deliver optimum conditions.		○	◎
7.4 Coordinate handover.	Production planning.		○	◎
7.5 Site acceptance testing.	Audit 7 and stage gate.	◎	○	○

FIGURE 2.6
EEM road map *improve* phase quality plan.

3. Technical stability.
   - Define the optimum conditions and the next improvement steps.
     - Summarize the results of the six steps; update the computerized maintenance management system (CMMS) and knowledge base.
     - Define steps to optimize the process.
     - Mobilize the *focused improvement* program.
   - Update technical documentation; track the project key performance indicator(s) and confirmation; capture the lessons learned.
4. Coordinate the handover.
   - Confirm the performance meets SAT criteria.
   - Confirm the tactics and resources for progression toward optimum conditions.

2.6.2 Step 8: Optimize

The optimization step involves the mobilization of the new asset-focused improvement program.

Although this may appear to be a theoretical step, it is characteristic of best-in-class organizations to maintain creative pressure on performance. These organizations have learned it is easier to maintain and improve performance than to recover from a deterioration of equipment condition in 12–18 months' time after performance has dipped, breakdowns are on the increase, and major restoration work is required.

2.7 SUMMARY

This chapter provides an overview of

- How the EEM road map integrates project delivery, project governance, and knowledge management
- The project governance learning landscapes
- The EEM organization and workstream RACI
- The responsibilities of management
- Setting out the project learning agenda
- EEM accountabilities for each step
- Other chapters covering these topics in more detail

Key messages include the following:

- Successful capital project delivery processes use structured decision phases that add detail at each step.
- Each phase covers a different decision landscape with different project governance challenges.
- EEM tools support these decision cascades throughout the complete process from concept to project delivery.

3

Design and Performance Management

Design and performance management is the first of the five EEM subsystems introduced in Chapter 2. This concerns

- The systematic definition, analysis, and evolution of design effectiveness
- Enhancing project value, reducing project risks, and minimizing total life cycle costs (LCCs)

There are close links between the topics here and those of Chapter 4 on specification and LCC management, which covers

- The capture of insights gained as a result of design management
- The use of document formats to aid the design process at each step—for example:
    - Models and sketches at the early stages encourage creativity and innovation.
    - Templates and charts aid analysis and option evaluation.
    - Drawings and layouts aid the development of material flows and workplace layouts.
    - Written documents are needed for contracts, risk assessment, project plans, tests, operational documentation, and training documents at later steps.

This chapter begins with an introduction to design management best practice before moving onto performance issues, avoiding design pitfalls, and details of the design activities during the define, design, and refine project phases introduced in Chapter 2, "The EEM Road Map."

3.1 IN SEARCH OF BETTER DESIGN

The purpose of the EEM design process is to achieve the lowest LCC. This approach is sometimes referred to as *design to life cycle costs* (DTLCC), an approach that extends the scope of design beyond the initial production day to include

- Early warning feedback of design weaknesses and opportunities to reduce LCCs prior to the beginning and following the end of the project.
- An analysis of the project option LCC drivers; this is then applied to design decisions. (LCC models are covered in Chapter 4, "Specification and LCC Management.")
- Post-day-one production design performance improvement through a systematic stepwise cycle of defining and achieving optimum conditions. This provides a structured learning process to enhance current designs and guide future capital projects.

Good design is achieved by a combination of innovation and systematic design processes. These processes provide a framework to apply value adding knowledge, guide learning, and trigger creativity. Like financial business processes, their value is enhanced by disciplined application.

3.1.1 Innovation

Innovation occurs most frequently by combining things that already exist to create new value and advantage. It is that "Aha!" moment when an idea suddenly clicks into place in a way that wasn't obvious at the outset.

The starting point for innovation is a trigger—something that gets the thought process going, lifts the veil of conventional thinking, and involves *gist-level* thinking to find new connections.

For example, a company at the start of their journey had decided to apply EEM to a project that involved squeezing a new production line into an available but rather small area. There was a concern that the material flow would need to involve external movement by forklift at night. Much work had been carried out by the project engineers on the shape and size of an airlock. During a *day in the life of* (DILO) review (see Section 3.2.4), the trigger of "how to move material at night" led the EEM core team to the idea of creating a new doorway into an adjacent building. In addition, they also suggested a way of improving the management of material flow that would reduce the amount of storage needed and release more space. There were some problems to overcome regarding differences in floor height, but the new suggestions were accepted. This innovation had a major impact on the success of the future operation, which broke all internal records for the pace of production ramp-up post day one.

An interesting phenomenon occurs with innovations such as this. With hindsight, it is often difficult to understand why any other option could have been considered. The result is that leaps in logic such as that in the example, with hindsight, can be seen as insignificant and not worthy of note. The reality is, however, that they are easy to miss. There are many examples of projects failures that hindsight suggests should have been easy to avoid but were not—the Hubble telescope, Terminal 5 start-up, the Channel Tunnel, and so on.

3.1.2 Systematic Design Development

For an interesting idea to be converted into something useful, *innovation* needs to be combined with a *systematic design development* process, similar to Edison's perspective: "Genius is 1% inspiration, 99% perspiration." It is no surprise, then, that the lack of a systematic design development process is what constrains many innovative and worthwhile ideas.

Good ideas can get lost in the general conversation if their relevance is not recognized or they are not yet fully formed. There is also a risk that group think leads to discounting useful ideas because they are not considered relevant. On the other hand, there will not be time to scrutinize every suggestion in minute detail. The matrix in Figure 3.1 illustrates how clarity of vision (what is required) and an understanding of cause and effect impact on conditions supporting the use of systematic thinking. When we are not clear about what we want to achieve and do not understand the cause–effect mechanisms involved we cannot use a systematic decision-making style. The matrix also illustrates why ideas are more likely to be overlooked by project leaders where the project vision is unclear. It also sets out how, if there is something we don't fully understand, as long as we have a collective view (in detail) of what we want to achieve, we can adopt a judgment-based approach to help us move toward systematic decision-making. If the vision is not clear and we are not sure about the cause–effect mechanisms, if we are lucky, we have a 50% chance of success.

In the preceding example, the project engineers' concern was how to get the equipment into a small space. They had limited knowledge about the day-to-day operation, so their choices were driven by that issue alone. The addition of a clear EEM vision for the project (Figure 3.2) and the inclusion of personnel with relevant front-line knowledge provided the trigger and context for the innovation to occur.

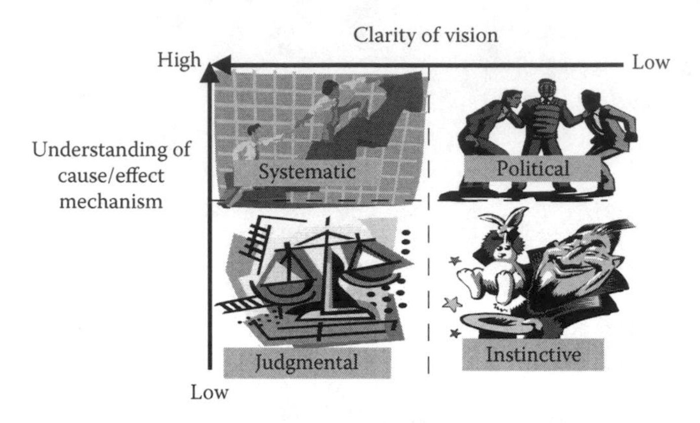

FIGURE 3.1
Decision-making styles.

Commercial	Operational	Technology
• Line speed targets achieved • Quality meets spec • Delivery on time • Early customer involvement • Flexible for new product formats. • Lightweight materials • Scope for future light weighting • Cost per case achieved	• Good access for asset care and cleaning • Line of sight visibility production • Familiarity with equipment • End-to-end line control • Visual management/Lean flow • Material handling equipment • Operator-led changeovers • Controlled work environment (noise, heat, light, etc.) • Integrated QC process to minimize time away from the line	• Clear maintenance plan • Ease of changeover • Safe compliance re equipment and environment for materials and people • Comfortable environment • Technical/engineering knowledge

FIGURE 3.2
Project vision: A trigger for innovation (case study).

The ability to apply Systematic design principles also depends on the clarity of available knowledge pools. A knowledge pool is an independent cluster of *explicit* and *tacit* knowledge. Explicit knowledge is contained in procedures and technical documentation and can be closed (not known to or accessible by us) or open (easily accessible when we need it). Tacit knowledge is knowledge we have without knowing it. It is difficult to write down and is typically only accessed under specific circumstances (Figure 3.3).

		Visibility	
		Closed	Open
Knowledge pools	Explicit	Core information available for use by authorized individual functions.	Information provided where and when needed through integrated systems and procedures including visual management.
	Tacit	"Blackbook"/closed information sources and unwritten rules.	Unstructured technical documents. Information available to those with knowledge of what to look for and where to look.

The role of knowledge management is to create explicit pools of knowledge.

FIGURE 3.3
Knowledge pools.

Often it is assumed that people jealously guard tacit knowledge, but mostly that is not the case. Most frequently, what is missing is a process to identify knowledge gaps and codify tacit knowledge so that it can be shared.

Knowledge management concerns the collation and codification of pools of knowledge in the form of

- Closed systems, which are needed to protect intellectual property or provide on-demand reference facilities
- Open systems, which provide information where and when it is needed—for example, visual management

Within the EEM process, tacit knowledge is structured under the six EEM design goals. They are as follows:

1. Customer quality
    - The process is able to meet current and likely future customer *quality*, *cost*, and *delivery* (QCD) features, including demand variability.
    - Provides a technology platform to support incremental product improvement.
2. LCC
    - The process has clearly defined cost and value drivers to support LCC reduction and maximize the return on capital invested.
3. Operability
    - The process is easy to start up, change over and sustain "normal" conditions.
    - Rapid close-down, cleaning, and routine asset care task completion.

4. Maintainability
   - Deterioration is easily measured and corrected.
   - Routine maintenance tasks are easy to perform and are carried out by internal personnel.
5. Intrinsic reliability
   - The function is immune to deterioration, requiring little or no intervention to secure consistent quality.
6. Intrinsic safety and environmental stability
   - The function is intrinsically safe, low risk, and fail safe.
   - Able to easily meet future statutory and environmental limits.

(The criteria for assessing the achievement of these design goals is included in Appendix A.)

These design goals describe the design parameters that have the biggest impact on LCCs. The capture of tacit knowledge is one of the outputs of EEM tools such as DILO and design module reviews.

Other knowledge management tools that support the translation of tacit knowledge into a design-friendly form include:

- Checklists
- Procedures
- Layouts and models

3.2 IN SEARCH OF BETTER PERFORMANCE

The challenges facing capital projects fit the definition of what are known as *wicked problems*—that is, a problem with many potentially conflicting design goals.* In this environment the combination of innovation and systematic design development are essential.

As any experienced designer will explain, the starting point for unpicking such complex problems is to gain an insight into the day-to-day reality of current users to guide their efforts.

* Horst Willhelm Jakob Rittel (1930–1990) was a design theorist and university professor who coined the term *wicked problem* to refer to a problem that is difficult or impossible to solve because of incomplete, contradictory, and changing requirements. These are often difficult to recognize. Also, because of their complexity, the effort to solve one aspect of a wicked problem may reveal or create other problems.

This is also the basis of the EEM principle of *maintenance prevention* (MP),* which is supported by the collation of MP data. Although the term may suggest a focus on maintenance engineering tasks, MP concerns the achievement of optimum performance with minimum human intervention by any function. In this book the terms *problem prevention* (PP) and *PP data* are used rather than MP to reflect the wider role of the principle. PP data is collated into *best-practice design books* and routinely updated to provide a resource to support future projects and improve current asset design. This is discussed in more detail in Chapter 4, "Specification and LCC Management," and Chapter 6, "Project Governance."

Four PP data capture and analysis tools support the delivery of systematic design. These are discussed in Sections 3.2.1–3.2.4 (Figure 3.4). These tools are used to support the achievement of

- Less operator intervention, including zero jams, one-touch/no-touch changeovers
- Predictable component life through the achievement of optimum component wear and the removal of sources of accelerated wear/human error
- Less quality assurance (QA) testing of the product because of confidence that the process is in control

Often the science behind the manufacturing process is not fully understood. As a result, the current generation of assets were developed based on the expectation that reliable output cannot be provided without frequent intervention and testing.

Operations personnel will have learned to live with the current level of process control. Technology personnel have learned to expect the need for batch process control or recycling loops. Commercial personnel are used to business-to-business customer terms and conditions that demand a high level of process checks. These "learned behaviors" can create inertia, which can make some feel that improvements to process control are not worth the effort.

In reality, to use a military expression, time spent in reconnaissance is seldom wasted. A relatively small effort to analyze and experiment with

* The term *maintenance* is the translation of a Japanese approach to the care and optimization of equipment performance through the joint efforts of all functions that impact on equipment effectiveness.

PP (Problem Prevention) data provides an insight into the shop floor reality to highlight failure modes, weak components and opportunities to prevent problems and defects. The outcome is to stabilize component life and extend time between intervention for operational routines including Production, Maintenance and Quality Assurance.

Tool	Purpose/Notes
Criticality assessment	Analysis of components by functional characteristic to identify weak components and provide a measure of design effectiveness. This includes critical to safety, reliability, operability, maintainability, customer value and life cycle cost mechanisms.
OEE and lifecycle cost analysis	Where are the main contributors to capitalcost, operational cost and OEE hidden losses. Data sources include Line history, run hours and maintenance cost data.
Condition appraisal	Assessment of basic conditions, identification of areas of accelerated deterioration. Used to identify refurbishment plans for assets which are not being replaced but which can impact on the delivery of flawless operation, Data sources include equipment standards, safety standards and breakdown analysis
"Day in the life of" (DILO)	Designed to tease out tacit knowledge about operational conditions and constraints as a trigger for setting standards, defining areas of risk, priorities and checklist points. Data sources include equipment trouble maps and line studies of planned and unplanned interventions.

FIGURE 3.4
PP data: Collating operational knowledge.

existing processes can provide a gateway to *step-out performance gains.**
For example, when a manufacturing company improved process control to increase the *mean time between interventions* (MTBI) from four minutes to more than two hours, this broke their historical link between the labor hours and equipment run hours. With this they were able to run equipment without a full-time operator in attendance.

The shop floor role changed from one of loading, overseeing, and unloading machines to one of managing supplier and customer order cycles. They still needed patrol inspections to monitor the equipment condition and performance, but performance was far more routine and predictable. The gains included

- Fewer defects and higher material yields
  - Material costs are often the largest contributor to manufacturing costs; improved process control has a significant impact on unit costs.
- Less time needed to achieve stable performance after a changeover
  - This means that smaller batches can be produced more easily. That in turn means increased flexibility to changes in customer demand, increased customer service, and less money tied up in finished goods.
- Less process stages, meaning a smaller plant footprint and lower capital costs

By incorporating the principle of PP into the design process some manufacturing plants have been able to achieve in excess of eight hours without intervention. Some process plants have achieved 12 hours. In these plants, the bulletin boards track days since the last defect in the same way that others report days since the last safety incident.

3.2.1 Criticality Assessment

A criticality assessment of the nearest current asset provides an insight into the strengths and weaknesses of the mechanisms/process controls against *functional characteristics*. The process provides the foundation data for codifying the tacit knowledge of those who know the asset best. It is also used to split the design problem into bite-sized design modules that are used throughout the project process.

* The term *step-out performance gains* refers to a multiple-factor performance gain that enables a new way of operating. This is in contrast with incremental improvement gains, which improve single-factor productivity gains such as labor efficiency or cost reduction.

Carry out the assessment with experienced front-line personnel and the EEM core team to identify what the core team currently know (or don't know) about the technology. Support this with breakdown data and root-cause analysis information where available.

Split the asset down into 10 to 15 main processes and assess each under the following eight functional characteristics (Table 3.1).

TABLE 3.1

Criticality Assessment Scales

Functional Characteristics	Assessment Basis	Rating	
Safety	If the module is in poor condition, what is the potential impact on safety risk?	High = 3	Low = 1
Availability	If the module is in poor condition, what is the potential impact on asset availability, including set-up?	High = 3	Low = 1
Performance	If the module is in poor condition, what is the impact on performance when running?	High = 3	Low = 1
Quality	If the module is in poor condition, what is the potential impact on the time taken to reach specification or defect levels?	High = 3	Low = 1
Reliability	Based on past experience, how reliable is this module (score 2 if unsure)?	Poor = 3	OK =1
Maintainability	If the module is in need of repair or servicing, how easy would it be to bring it into working order?	Hard = 3	Easy = 1
Environment	If the module is in poor condition, what impact will this have on the environment (local and external)?	High = 3	Low = 1
Cost	If this module needs repair, what is the likely cost, including materials, labor, and lost production?	High = 3	Low = 1

Record each assessment and total the scores across the columns.

The following is a criticality assessment example for a filling line (Figure 3.5).

The analysis in Figure 3.5 is sorted by the total criticality score, but the most interesting column is *R* (reliability). EEM defines *reliability* as a measure of intervention.

This means that, for example, in the case of a filter, even if it is functioning correctly, if it needs frequent cleaning it is classified as having low reliability. The assessment is of the reliability of the system. Those areas assessed at a score of 2 or 3 can then be targeted to guide the design process towards an asset which requires minimal routine attention.

Case. prelim | Crit. above 14

	S	A	P	Q	R	M	E	C	TOT
Process supply	2	3	3	3	2	1	3	3	20
Services	2	3	2	3	2	2	3	3	20
Pasteurizer	3	3	3	3	1	1	2	2	18
Velcorin dosing unit	3	3	1	3	1	2	3	1	17
Sleever	1	3	2	3	2	2	1	2	16
Material	1	2	2	3	2	1	2	2	15
Labeler	1	2	3	3	2	1	1	2	15
Carbonate	1	3	3	3	1	1	1	1	14
Blow moulder	1	2	3	2	2	1	1	2	14
Filler	2	2	2	2	2	1	1	2	14
Palletizer	2	2	2	3	2	1	1	1	14
Multipack	1	2	2	3	2	1	1	1	13
Case packer	1	2	2	3	2	1	1	1	13
Coding	1	3	1	3	1	1	1	1	12
Material handling	2	1	1	2	1	2	2	1	12
Capper	1	2	2	3	1	1	1	1	12
Bottle inspection	2	1	1	3	1	2	1	1	12
AGV	1	2	2	1	2	2	1	1	12
Mes	1	3	1	2	1	2	1	1	12
Spiral	1	1	2	1	2	1	1	2	11
Bar code label	1	1	1	3	2	1	1	1	11
Dryer	1	1	2	2	1	1	1	1	10
Pallet wrapper	1	1	1	2	1	1	1	1	9
Conveyor to palletizer	1	1	1	1	1	1	1	1	8
Average	1.42	2.04	1.88	2.50	2.00	1.29	1.38	1.46	

Rating against benchmark

Asset care priorities		
A	B	C
E		
E		
S		
S		
Q		
Q		
Q		
	R	
	R	
Q		
Q		
Q		
Q		
		X
Q		
Q		
	R	
Q		
		X
		X

S+R+M	A+P+R	P+Q+R	M+R+C	Q+R+C	C+R+E
Safety risk	Relability	Operability	Maintainability	Cust. value	LCC
5	8	8	6	8	8
6	7	7	7	8	8
5	7	7	4	6	5
6	5	5	4	5	5
5	7	7	6	7	5
4	6	7	5	7	6
4	7	8	5	7	5
3	7	7	3	5	3
4	7	7	5	6	5
5	6	6	5	6	5
5	6	7	4	6	4
4	6	7	4	6	4
4	6	7	4	6	4
3	5	5	3	5	3
5	3	4	4	4	4
3	5	6	3	5	3
5	3	5	4	5	3
5	6	5	5	5	4
4	5	4	4	4	3
4	5	5	5	5	5
4	4	6	4	6	4
3	4	5	3	4	3
3	3	4	3	4	3
3	3	3	3	4	3
4.25	5.46	5.92	4.29	5.50	4.38
95%	71%	62%	94%	70%	93%

FIGURE 3.5
Criticality assessment example.

The combination of reliability and other factors helps us to assess design weaknesses using the following six EEM design goals:

- Intrinsic safety (safety + reliability + maintainability)
- Intrinsic reliability (reliability + availability + performance)
- Operability (performance + reliability + quality)
- Maintainability (maintainability + reliability + cost)
- Customer value (quality + reliability + cost)
- LCCs (environmental loss + reliability + cost)

Each combination of assessments could score a maximum of 9. Anything above a 6 is considered a weak component. Those scoring 8 or 9 have the most significance. A number of other insights are provided by the analysis—for example:

- The profile of the process supply, services, and labeler indicate that these functions have high complexity. These are areas that would benefit from simplification.
- Failure modes.
  - Functions with a 3 in the *availability* column are subject to sudden failure.
  - Functions with a 3 in the *performance* column have a high impact on throughput rates.
  - Functions with a 3 in the *quality* column contribute most to quality defects.
  - Functions with a 3 in the *environmental* column are prone to leaks when damaged, and many need protection from damage/bunded areas.

The final step of the criticality analysis is to identify which conditions would be needed for the process to be optimized—for example, conditions such as being dry as well as leak and dust free. These condition standards can then be used during condition appraisal activity to assess the current equipment condition and identify sources of accelerated wear.*

Later in the project, the criticality assessment tool is also used to assess the new asset/process to support the development of an *asset care* plan. This is discussed further in Chapter 4 (Figure 3.6).

---

* Component wear patterns can be categorized into normal wear and accelerated wear; for example, the corrosion of electrical contacts due to an ingress of water is accelerated wear. Electrical contacts should be protected from water. Similarly, failure due to a lack of or over-lubrication is also accelerated wear. An analysis of failure data from over 500 years of run time history identifies that around 20% of component failures are premature and due to accelerated wear. A further 45% of failures are due to the lack of replacement when worn.

Top-level assessment to identify risk and complexity

	S	A	P	Q	R	M	E	C	TOT
Process supply	2	3	3	3	2	1	3	3	20
Services	2	3	2	3	2	2	3	3	20
Pasteurizer	3	3	3	3	1	1	2	2	18
Velcorin dosing unit	3	3	1	3	1	2	3	1	17
Sleever	1	3	2	3	2	2	1	2	16
Material	1	2	2	3	2	1	2	2	15
Labeler	1	2	3	3	2	1	1	2	15
Carbonate	1	3	3	3	1	1	1	1	14
Blow moulder	1	2	3	2	2	1	1	2	14
Filler	2	2	2	2	2	1	1	2	14
Palletizer	2	2	2	3	2	1	1	1	14
Multipack	1	2	2	3	[illegible]	1	1	1	13
Case packer	1	2	2	3	2	1	1	1	13

Sub modules assessment to develop asset care plans

Palletizer	R	TCT	Priority A	Priority B	Priority C
Low-pressure air	1	10			×
Dividers	2	14		R	
Table	1	9			×
Layer pad arm	2	12		R	
Empty pallet feed	1	8			
Pallet rollers	1	[illegible]			×
Twister box	1	11			×

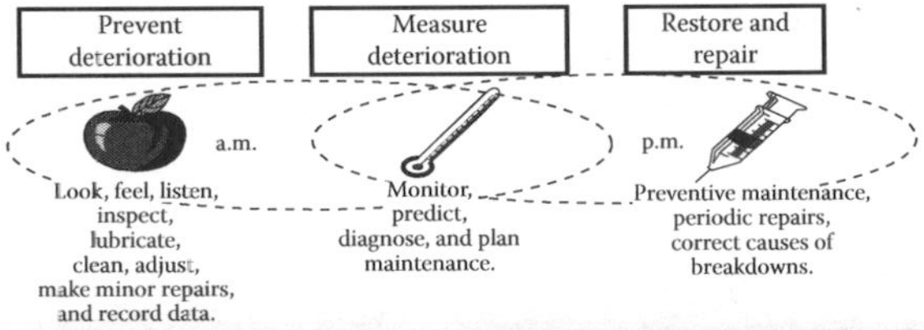

Prevent deterioration — a.m. — Look, feel, listen, inspect, lubricate, clean, adjust, make minor repairs, and record data.

Measure deterioration — Monitor, predict, diagnose, and plan maintenance.

Restore and repair — p.m. — Preventive maintenance, periodic repairs, correct causes of breakdowns.

Prevent deterioration		Measure deterioration				Restore deterioration				
Contamination control/deterioration check	Lubrication task	Tasks to confirm normal conditions	Tasks to detect hidden failiure	Test for minor quality defects	Test to identify onset of failure	Simple parts replacement task	Adjust for normal wear	Restore correct tightness	Overhaul	Replace
1	1					1	1		1	
1	1	1	1	1	1		1	1	1	1
1	1		1	1	1		1		1	
	1	1	1	1	1		1		1	1
1	1		1				1		1	
1	1	1	1		1	1	1	1	1	1
1	1	1	1	1	1	1	1	1	1	1

FIGURE 3.6
Criticality assessment and asset care plans.

3.2.2 Hidden Loss Analysis/Trouble Map

From the equipment history and discussions with experienced personnel, produce a list of common problems categorized using the following six loss headings. Include all areas of unplanned, ad hoc intervention such as additional handling or just-in-case inspection. Try to account for the root causes of effectiveness losses using key performance data such as an *overall equipment effectiveness* (OEE) report (Table 3.2).

TABLE 3.2

Defining the Hidden-Loss Trouble Map

Hidden Loss	Notes	Typical Countermeasure
Breakdowns	Failure due to deterioration or failure of a component	Raise equipment condition and improve early detection.
Setup and adjustment time	Activities to prepare the equipment/process for operation (including preparation, cleaning, and unplanned adjustments)	Organize tasks to minimize downtime (simplify, combine, eliminate).
Idling and minor stops	Stoppages such as trips, blockages, and unplanned interventions; typically less than five minutes in duration	Establish optimum process conditions.
Speed losses	Lost capacity due to running at a speed lower than theoretical maximum	Address quality issues to release capacity.
Start-up material losses	Out-of-spec product produced prior to achieving steady-state operation (and following close-down)	Optimize process preparation to reduce time to good material.
Rework and quality defects	Material processed more than once to achieve specification	Optimize running conditions.

Quantify the frequency or relative frequency of each loss area. If necessary, introduce a short-term data capture program to identify priorities for action including.

- Known performance problems and hidden losses for each design module
- A prioritized list of actions to deal with current issues, such as
  - Equipment conditions
  - Working methods
  - Design weaknesses
- Data capture and analysis tasks to confirm the levels of each problem area and their impact on project deliverables (Figure 3.7)

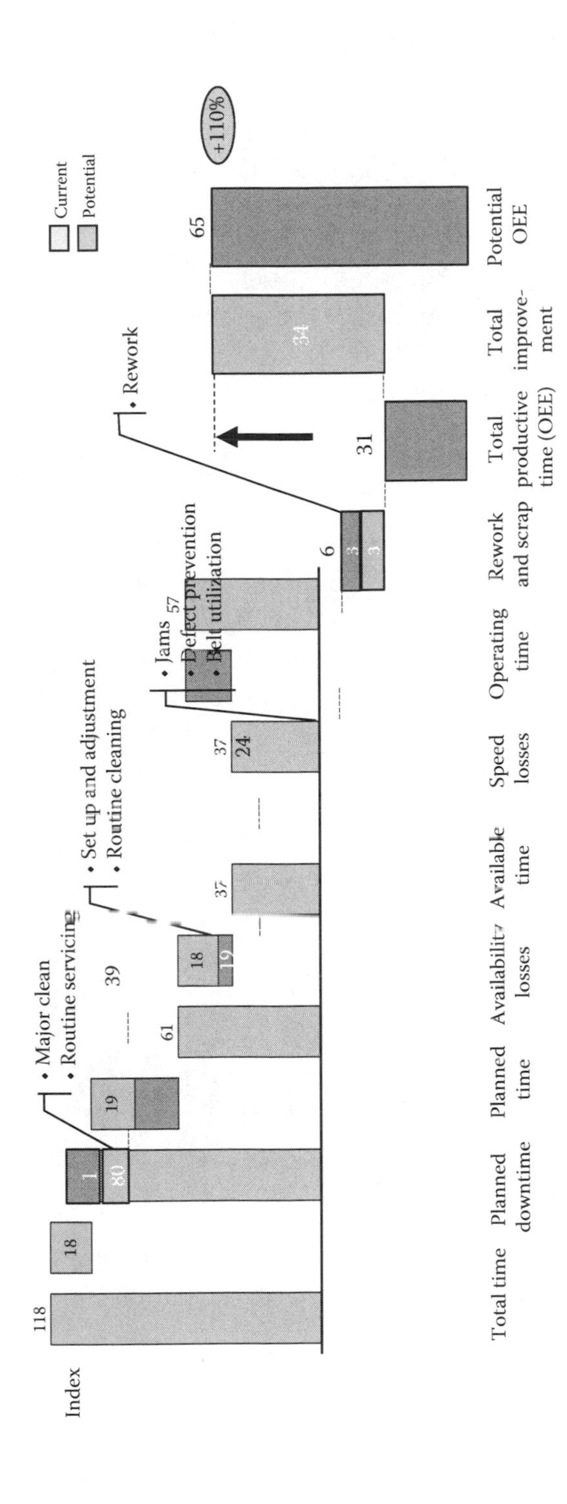

FIGURE 3.7
Understanding hidden losses and potential gains.

3.2.3 Condition Appraisal

Inspect every square centimeter for deterioration from the ideal standard identified as part of the asset criticality review. Look for

- Dirty, damaged, or neglected mechanisms or hoses
- Missing nuts and bolts
- Leaks

Identify sources of accelerated wear such as scattered dust and dirt. Pay particular attention to high technical stability risk areas from the criticality assessment. Take photographs where practical to record the problem area.

Aim to inspect critical areas most thoroughly. This may require in-depth cleaning or strip-down. The cleaning will help to identify difficult-to-clean areas and sources of contamination, raise understanding of the asset, and set a visual standard for equipment conditions.

Identify any obvious low-cost improvements, such as visual indicators to reduce the risk of error. Review the refurbishment actions and ideas generated and categorize them by high, medium, and low costs and high/low benefits.

Prioritize the low-cost, high-benefit items for immediate attention, even where the asset is to be replaced. Low-cost, high-benefit items will often recover their cost in weeks if not days. The evidence of the gains will also be a useful lesson learned about the importance of looking after the new asset once it is in place.

3.2.4 DILO Reviews

The DILO review is a tool for

- Collating knowledge about the commercial, operational, and technology aspects of the new operation
- Gaining an insight into the use of a new asset and how to deliver its full potential
- Understanding the scope of actions needed to deliver flawless operation from day one

Table 3.3 sets out a number of ways in which the DILO process can be used.

TABLE 3.3

Uses of a DILO Review

Purpose	Goal	Outputs
Communication	To bring all parties up to speed with ideas and confirm design effectiveness/gaps	Benefits, concerns, next steps
Assessment of design or working methods	To evaluate the proposed design/ working method compared with EEM guidelines and standards	List of agreements reached/areas for attention
Identify barriers to flawless operation	To summarize the issues raised under design goals, guidelines, and standard categories to support the design process	Checklist of issues to address within the review process
Identifying improvement opportunities	To generate ideas for improvement and identify the most attractive	List of improvement ideas and potential

The DILO review involves a simple desktop or even full-scale, hands-on simulation process using

- A physical representation of the process under review (e.g., a layout drawing, floor markings, or a cardboard model)
- A next-event list/script of activities to be simulated—for example, those required during start-up, steady state, and close-down of the process under review (Figure 3.8)

Use a physical representation of the process under review rather than a process flow sheet or process and instrumentation drawing. Also match the level of detail on the drawing to the EEM step requirements: simple outlines at the concept step and more detailed drawings at the detailed design step.

1. Involve people who understand the day-to-day reality of the process under review. Invite participants covering commercial/customer-facing, operations, and technology perspectives.
2. Explain the review goal and planned outcomes and allocate each person an EEM goal to consider (safety, reliability, operability, maintainability, customer value, LCCs). A summary of each goal is included in Appendix A.
3. Step through the next-event list using the layout/model to help participants visualize the physical context.

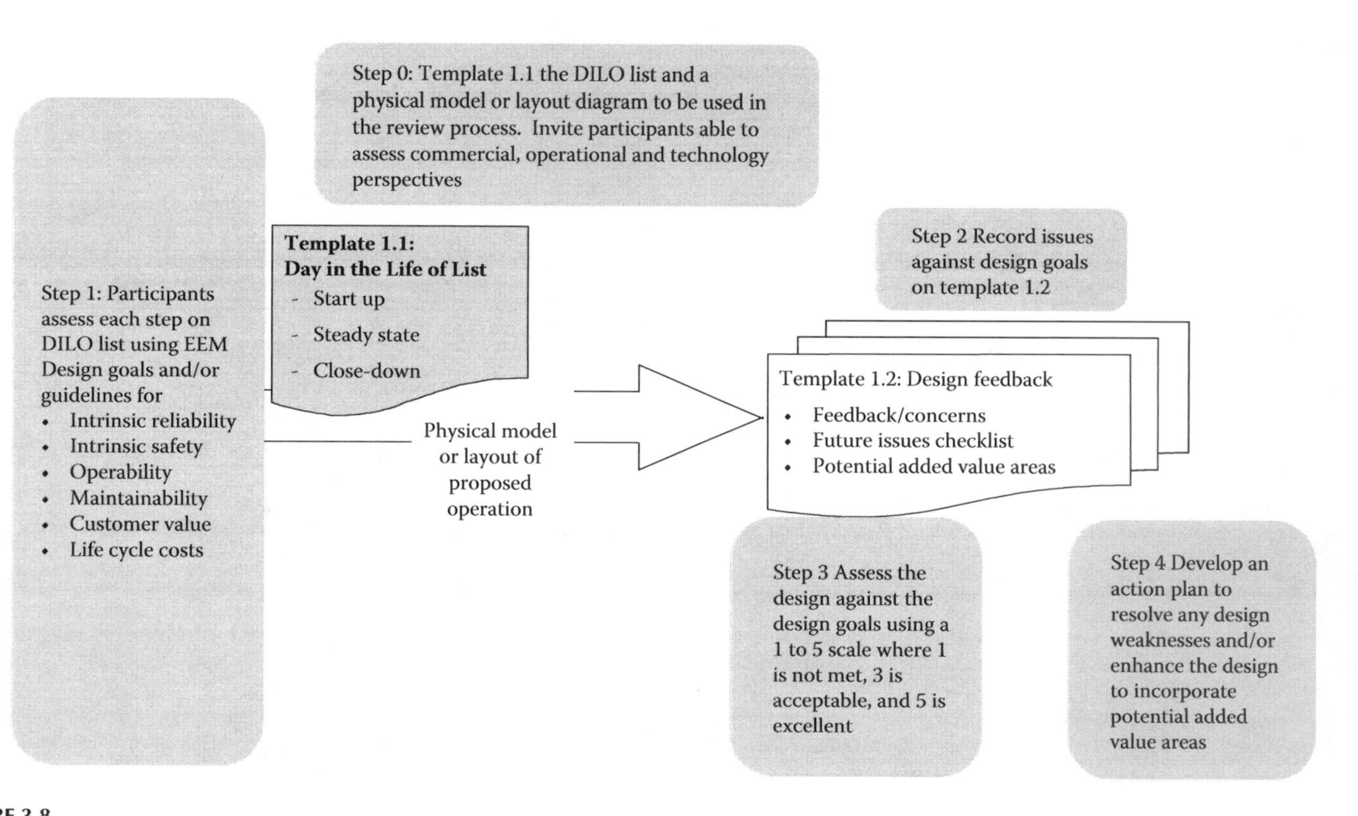

FIGURE 3.8
DILO simulation.

4. During the review event ask participants to capture issues that have the potential to impact the delivery of flawless operation. Cluster the points under common headings to be explored later.
5. Toward the end of the session involve participants in collating common themes to develop standards and checklist points to support future EEM road maps.
6. Develop action plans to advance the issues raised.

The following is an example of a next-event list from a pasteurizer DILO review.

A. Start-up

1. Close tank doors (10 each side doors)
   Multiple ratchet fasteners per door
2. Start operating sequence
   Discuss current purified water supply issues
   Fill with blend of RO water
   Takes 20 minutes to flood zones
3. Check that water has reached required level and has not timed out
   Visual check of levels at multiple locations
4. Turn steam and air supply and pumps on (multiple locations)
5. Leave until temperature achieved (45 minutes to 1 hour)
6. Not sure how and where we handle
   Dosing
   Acid addition

B. Steady state

7. Red post checks every three hours
8. DIP slides
   Bio count
   Sample point design issues (improved from standard approach, should we keep this?)
9. Alarms
   On over-temp pasteurizer, crash cools stopping process for around five minutes Under temp
   Operator investigates problem to find out source of problem and raise technician
10. Replace chart recorder every 24 hours

C. Shut-down

11. Run out pasteurizer
    Open doors, push cans through to clear line
12. Turn off steam
13. Let pasteurizer cool
    Lift lids, scrub sieves, isolate air
14. To gantry, turn off pumps
15. Downstairs, open drain valves to reduce levels
    Open doors to remove final water content (won't drain completely without this)
16. Inspect for growths

3.3 AVOIDING DESIGN PITFALLS

3.3.1 Creating Design Modules

During EEM step 2, *high-level design*, split the design scope into six to eight design modules. For larger projects these design modules can be further divided into submodules. Apply the design module structure throughout the project to support value engineering, tender document design, vendor selection, detailed design, installation, and commissioning.

The benefits of the design module approach include

- *Improved ease of project visualization*: Providing a bridge between top-level project governance and detailed level tasks. This aids progress tracking and understanding the interdependencies between modules.
- *Value engineering*: The functional description of each module combined with definitions of inputs and outputs provides the information to assess where value is added and costs can be removed.
    - Value engineering includes the use of simple analysis tools such as X-charts, Lean mapping, and OEE analysis to explore cause–effect mechanisms. This includes how upstream processes impact on downstream performance for both external and internal links in the value chain.
- *Vendor collaboration*: Each tender package is set out in terms of the delivery of a common set of targets, time lines, and construction

criteria for all vendors at a design module level. This aids project planning, progress management, and the resolution of compatibility issues.

- *Project governance*: The module structure provides a framework to prioritize critical areas and manage the progress of project deliverables at a common level of detail.
    - Progress reporting is carried out for each module at every step to avoid the risk that equipment areas are overlooked or remain hidden within a larger design specification.
    - Specification, procurement, design assessment, witnessed inspection/testing criteria, training, and technical documentation management are linked together, using a common framework throughout the project.

Figure 3.9 is an example of a functional specification made up of eight design modules, each containing a transformation processes step. The color version of the format includes red, amber, and green color coding to highlight the relative level of criticality from the criticality assessment (see Section 3.2.1).

It is worth noting that in the previous example, the most critical module (material supply and process blending) was an existing facility and not part of the capital budget. This illustrates the importance of considering a complete end-to-end process.

3.3.2 Incorporating PP Data into Design

The collation and analysis of PP data provides the insight to define design modules, assess relative module criticalities, and identify points of interest within them. Figure 3.10 illustrates how this information is then used throughout the EEM road map to confirm the efficacy of actions to address the causes of defects and problems, and to assure progress through the fabrication, installation, and commissioning steps.

3.3.3 Recognizing Vendor Skill Sets

Every reputable vendor wants to do a quality job. They will want to have your installation as a great reference site so that they can impress future potential customers with the value of their work. You will want the same.

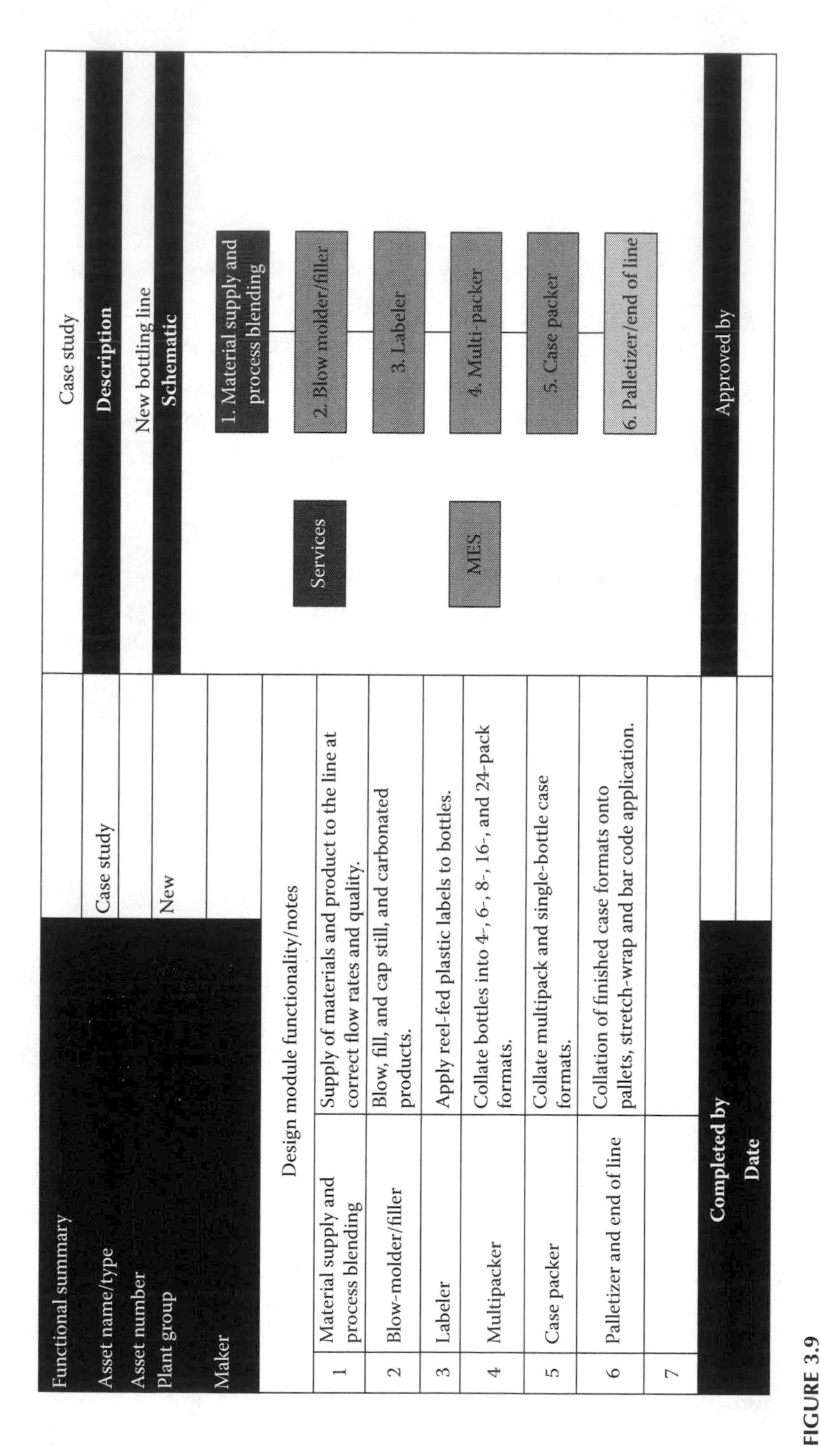

Functional summary			Case study
Asset name/type		Case study	**Description**
Asset number			New bottling line
Plant group		New	**Schematic**
Maker			
	Design module functionality/notes		
1	Material supply and process blending	Supply of materials and product to the line at correct flow rates and quality.	
2	Blow-molder/filler	Blow, fill, and cap still, and carbonated products.	
3	Labeler	Apply reel-fed plastic labels to bottles.	
4	Multipacker	Collate bottles into 4-, 6-, 8-, 16-, and 24-pack formats.	
5	Case packer	Collate multipack and single-bottle case formats.	
6	Palletizer and end of line	Collation of finished case formats onto pallets, stretch-wrap and bar code application.	
7			
Completed by			Approved by
Date			

FIGURE 3.9
Functional definition/design module summary.

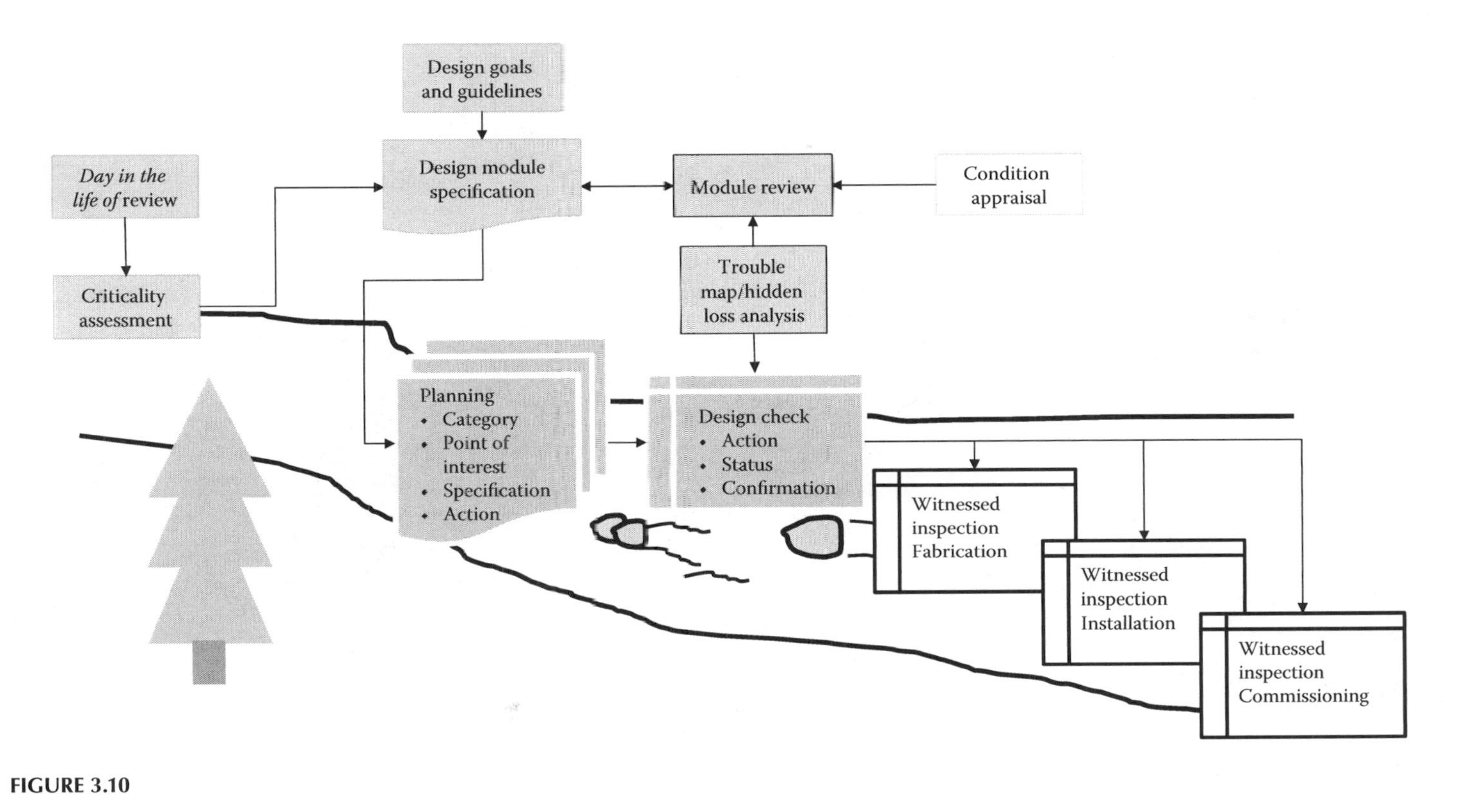

FIGURE 3.10
Problem prevention design flow.

It is important to recognize that vendors cannot directly influence all of the factors that impact on the delivery of flawless operation. There are a number of commonly held but unreasonable expectations of vendors that can create tension in working relationships and ultimately impact on project deliverables.

Common pitfall 1: Assuming that the equipment vendor understands your business.

In general, the people that design equipment have limited experience of using the equipment they design. Major exceptions to this are manufacturers of consumer goods such as automobiles and cell phones. The design of these items has been consistently improved over time in part because those who make the products also use them. This is not true of manufacturers that make equipment for other manufacturers to use.

Naturally, vendors will understand the functions of their equipment, but not necessarily how it is best used. The only way vendors would have a similar level of knowledge of day-to-day operations is if they were competitors. In which case, they are unlikely to be your preferred vendor.

This lack of firsthand experience of using equipment can lead to designs that are difficult to operate and prone to human error. This can lead to design weaknesses that are the equivalent of car designers placing features such a level gauge next to the petrol tank rather than in the driver's line of sight.

Common pitfall 2: Assuming that the sales team and the internal vendor design team are a single entity. As with most organizations, these will be different departments with different people, outlooks, and priorities.

Common pitfall 3: Ignoring the reality that vendors can only offer you what they have, rather than what you may need.

Common pitfall 4: Assuming that vendor training is all that is needed to develop the internal skills and capability needed to use the new asset.

The best way to help a vendor to achieve an outstanding result is to know what you need, set clear expectations, communicate these to the vendor, and work with them as equal partners to deliver the best possible project value.

An oil and gas exploration company assessed that the additional year-one gains made by achieving flawless operation from day one were worth as much as the capital cost. They also understood that this would not have been achieved without close collaboration with the vendor based on a cross-company teamwork built on a clear understanding of strengths and weaknesses.

Naturally, vendors will be held to account for the products and services they agree to deliver, but a relationship based on collaboration delivers major benefits, including

- Improvements from an *average* to a *simplified* design, worth a 30% reduction in capital costs
- Improvements in operability and maintainability, reducing running costs by as much as 50%

Even where the options to influence the design are limited, the improved understanding gained by working closely with vendors make this an important feature of the journey to flawless operation. This can have a major impact on the design of the new operation, the workflow, skills needed, and so on.

3.3.4 Understanding Behavioral Bias

As mentioned in Chapter 1, at the early steps of a project, decisions have to be taken with little quantitative data. The available quantitative data is often of a technical nature, which requires interpretation, or of a financial nature, which is often too simplistic to take into account the full range of cost drivers.

The use of such data can lead to the use of inbuilt behavioral weaknesses/paradigms* such as *framing*.†

Consider the decision made by a taxi driver who defines a target level of earnings needed each day. He decides when he has reached the level of income at which he can end his shift, safe in the knowledge that he has earned enough to pay the bills. This seems a logical commonsense approach that we can all understand.

However, if you widen the frame of reference you discover that when it is raining more people want to take cabs and the target may be achieved by midday, but on days when it is sunny and people prefer to walk the target may not be achieved at all. Wouldn't the driver be better off working when

* Behavioral economics research has identified a number of behavioral models to explain the mechanisms that drive decisions and choices that appear to defy conventional logic.

† Erving Goffman, *Frame Analysis: An Essay on the Organization of Experience*, Harvard University Press, Cambridge, MA, 1974.

it rains and heading for the beach early on sunny days? Now the original logic doesn't make sense. This illustrates how logic and decision-making is influenced by the frame of reference.

The project engineers working on the installation of a production line in a small area in the earlier case study were also influenced by framing. Their concern was how to get the equipment into a small space. They had limited knowledge about the day-to-day operation, so their choices were driven by that issue alone. The development of a comprehensive project vision widened the frame of reference and led to the inclusion of personnel with relevant front-line knowledge. Once the frame of reference was widened, the improved solution seemed obvious.

Other behavioral weaknesses that can bias decisions when quantitative data is not available include the use of untested rules of thumb, a natural tendency to be overconfident about future abilities, and group think. There is also evidence of the presence of each of these in the earlier case study.

In this environment the risk is that decisions are made based on *highest-paid personal opinion* (HiPPO). Even if the decisions taken by that person are correct, to be able to deliver the full potential, those involved need to understand why that decision was the right one.

3.3.5 Guiding Design Decisions

What is needed is a decision-making process to

- Set out the full scope of issues to consider
- Challenge weak thinking
- Develop insight into areas of uncertainty even when quantified data is not available

We also need a decision-making process that supports design-related tasks throughout the concept to beneficial operation, including

- Concept option development
- High-level design option evaluation and risk reduction
- Vendor selection
- Equipment and operational method design and simplification
- Checklist and test development
- LCC reduction (Figure 3.11)

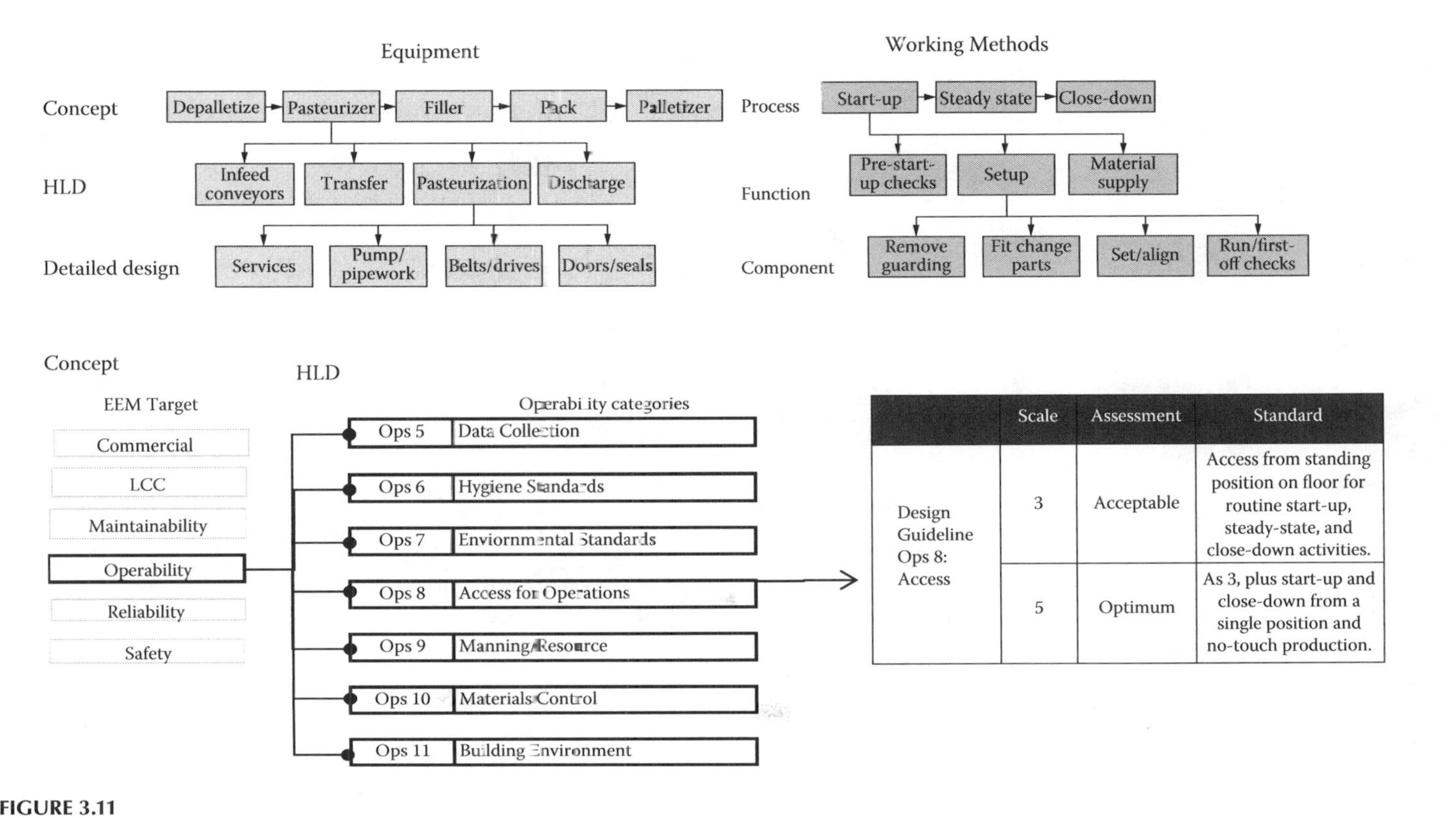

	Scale	Assessment	Standard
Design Guideline Ops 8: Access	3	Acceptable	Access from standing position on floor for routine start-up, steady-state, and close-down activities.
	5	Optimum	As 3, plus start-up and close-down from a single position and no-touch production.

FIGURE 3.11
Design evolution: Adding detail at each step.

EEM overcomes these behavioral pitfalls by applying a hierarchy of design goals, guidelines and standards to match the increased granularity of design decisions at each step of the EEM road map (Figure 3.12).

3.3.6 Design Goals

At the top of the EEM Heirachy are the 6 EEM design goals. These 6 areas have a major impact on life cycle costs and are used throughout the EEM process. Below they are used to set out the principles of low life cycle cost equipment design.

- *Customer quality*: The process is able to meet current and likely future customer QCD features and demand variability, and provides a platform for incremental product improvement.
- *Life cycle cost*: The process has clearly defined cost and value drivers to support LCC reduction, enhance project value, and maximize the return on capital invested.
- *Operability*: It is easy to start up and change over the process and to sustain "normal" conditions. Rapid close-down, cleaning and routine asset care task completion.
- *Maintainability*: Deterioration is easily measured and corrected. Routine maintenance tasks are easy to perform and are carried out by internal personnel.
- *Intrinsic reliability*: The function is resistant to deterioration, requiring little or no intervention to secure consistent quality.
- *Intrinsic safety and environmental stability*: The function is intrinsically safe, low risk, fail safe, and able to easily meet future statutory and environmental limits.

In Appendix A, these design goals have been calibrated against a 1–5 scale, where 1 is poor/limited, 3 is acceptable, 4 is a history of measurable improvement, and 5 is a significant history of measurable improvement. This scale was developed for a food and drink manufacturer. For your own industry, define what a score of 3 (acceptable) looks like. Treat this as a minimum. Reserve a score of 4 for exceptional and 5 for higher than that.

3.3.7 Design Guidelines

The second tier of the heirarchy comprises of a set of design guidelines. These are process specific and are used to support *invitation to tender*

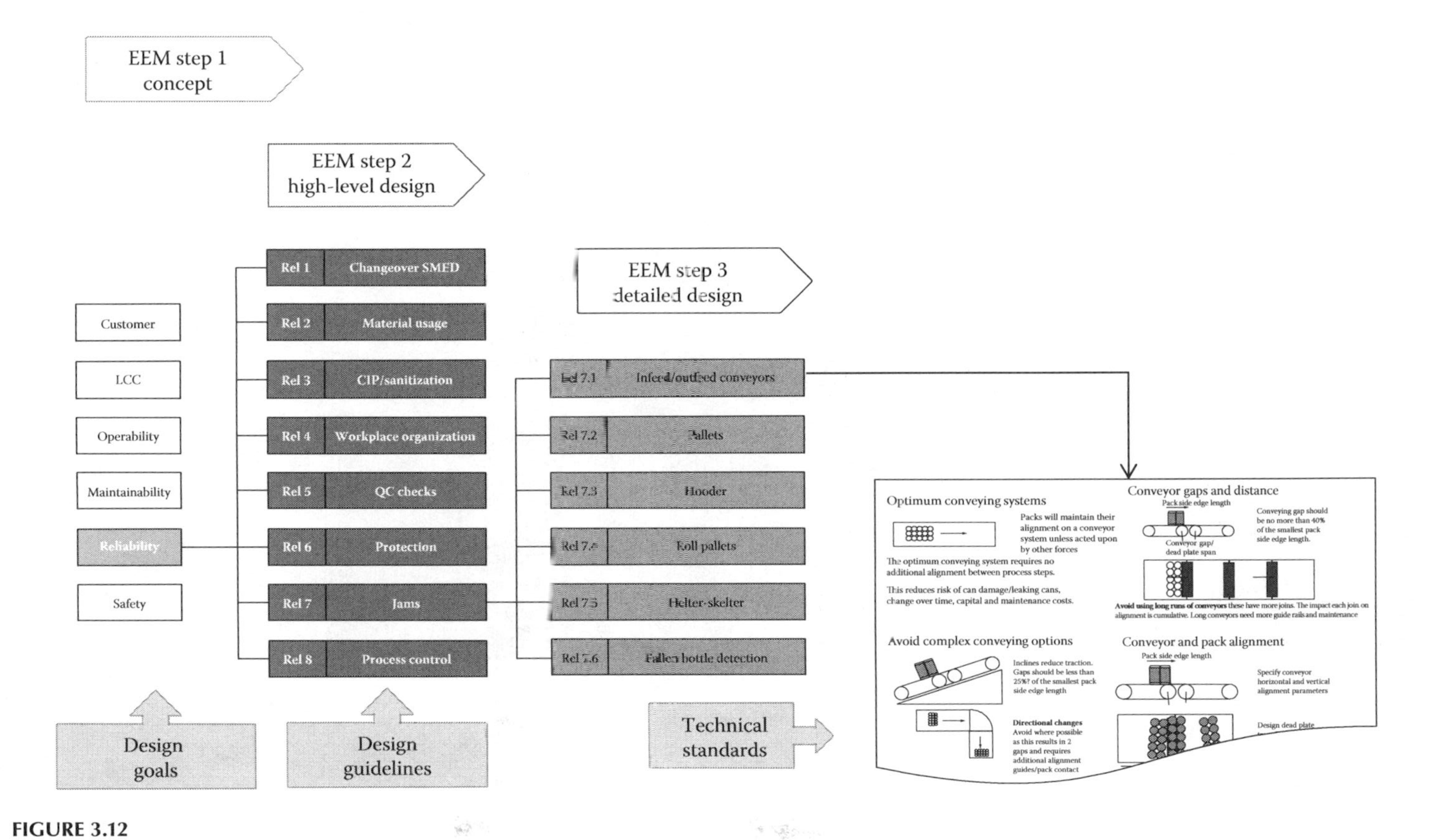

FIGURE 3.12
EEM design goal, guidelines, and standard hierarchy.

(ITT) or *request for quotation* (RFQ) documents, and detailed design and risk assessment activities.

Ideally, limit the number of design guideline categories to around 10 per design goal. The more guideline categories, the longer each design assessment takes. Consider the guidelines as a scorecard for design assessment and an agenda for discussion rather than a detailed specification. Ten categories per design goal results in a 60-point assessment. This should be sufficient to identify the strengths and weaknesses of design options and support the evaluation process.

Sources of data to support the development of these design guidelines include

- DILO reviews
- Module reviews
- Technical literature, operating manuals, and catalogues
- PP data
- Equipment history and line reports
- Root cause analysis and focused improvement activities
- Accident reports

When defining the categories, set out the reasons for defining that category to avoid duplication. An example of design guideline categories is set out in the Table 3.4.

For each category define the assessment basis against a 1–5 scale, where 3 is acceptable and 5 is optimum.

Figure 3.13 is an example of acceptable and optimum characteristics for the preceding design guidelines.

These examples are from a company at the early stages of implementing EEM. As can be seen, guidelines for an assessment of 3 (acceptable) have been defined, but there are gaps in the definition of 5 (optimum). It is not unusual for these higher-level guidelines to be set after experiencing multiple EEM projects.

3.3.8 Technical Standards

These are more detailed and cover component-level criteria. They support the standardization of components, build details, and therefore, working methods. This includes areas such as wiring colors, labeling, and spare

TABLE 3.4

EEM Operability Design Guidelines (Case Study)

	Category	Scope	Issues/Weakness
Ops1	Data collection	Production data: collection and analysis	Data capture quantity and method to meet relevant company standards and stakeholder needs, combining data from multiple sources into the required format, real-time data needs, exception reports/flags, and response/decision-making.
Ops2	Hygiene standards	The elimination of sources of dirt, and the ability to maintain clean and hygienic equipment	Compliance to CIP site microbiological standards, aseptic vessels and mains, sample point contamination, resistance to pressure cleaning and foam/chemicals, shadow areas.
Ops3	Environmental standards	Environmental management and loss control	Map, monitor, and minimize all material, process, and packaging waste flows; assure compliance to ISO14001.
Ops4	Access for operations	Access for operability and difficult activities	Low-level access, designated material storage areas, room to maneuver; identify complex or difficult start-up, steady-state, close-down, and cleaning tasks.
Ops5	Materials control	Material flow-in; start-up, running, close-down conditions	QA protocol; changeover, including workstream automation and watchdog monitoring.
Ops6	Building environment	Building and facilities design standards	Ventilation, floor, walls, ceilings, lighting, access, and drainage.

parts provision, which have a major impact on training time, human error, and operational costs (Figure 3.14).

As with the more generic design guidelines, their definition should include

- The reasons for preparing the standard
- The scope of the standard
- How the standard is to be applied

Categories			Scope		3. Acceptable	5. Optimum
Operability	Ops 1	Data collection	Production data-collection and analysis	1.1	Provision of data capture, analysis, and reporting for each stakeholder which meets BCS standards	Automatic data capture and by exception reporting
	Ops 2	Hygiene standards	Elimination of sources of dirt, and ability to maintain clean and hygienic equipment	2.1	All plant installed should be resistant to detergents used for foam cleaning and pressure washing, or alternative external clean procedure must be provided	
				2.2	Where applicable, equipment must meet FEMAS standards for bi-products	
	Ops 3	Environmental standards	Environmental management and loss control	3.1	Target zero losses by defining clear waste routes, monitored flow, containment where necessary, and intrinsic minimum waste within design	Zero losses/self-sufficient site, minimal waste leaving site
	Ops 4	Access for operations	Access for operability and difficult to do activities	4.1	Low-level and unrestricted access from standing position on floor to operating machine for the following conditions: start-up, steady running state, and shutdown	All routine operations tasks to be easily achieved without stopping the process
				4.2	Where working at floor level is not possible, appropriate platforms and lifting equipment are incorporated in the design (fixed, mobile, specific) to allow safe access to/and movement of equipment where needed	
				4,3	No difficult to clean places or components. All areas that require sampling for cleanliness are easy to access	
	Ops 5	Materials control	Material flow in; start-up, running, shutdown conditions	5.1	Quality analysis met using in-line monitoring and exception alarm reporting	Fully automated change over/self-monitored plant
				5.2	Eliminate, where reasonably practicable, the need for manual handling of materials. Access to materials unrestricted at point of use	
	Ops 6	Building environment	Building and facilities design standards	6.1	Clean and dry floor operation by defining clear standards for elimination of sources of contamination, including drainage (floor and other process drains), ventilation, aspiration, conveying, steam and liquids, and materials waste and sampling	

FIGURE 3.13
Design guidelines example.

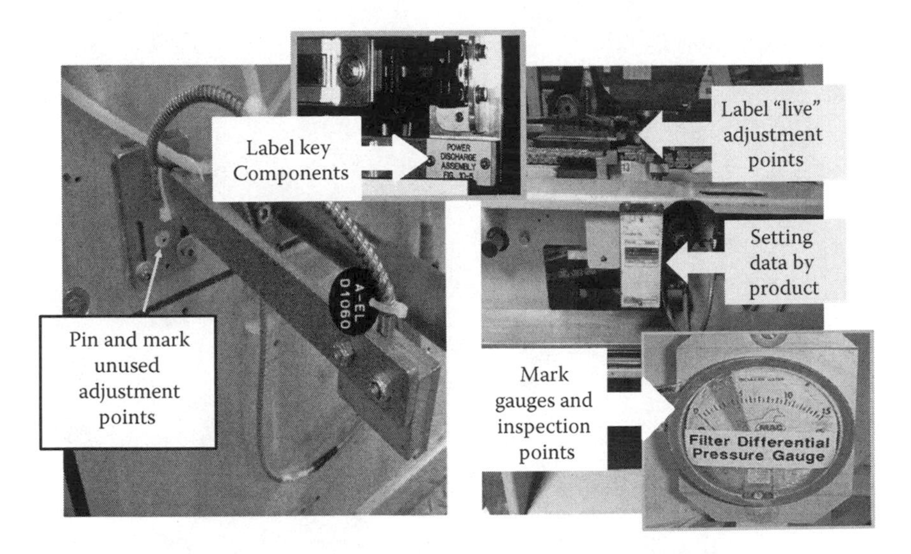

FIGURE 3.14
Setup and adjustment technical standards.

It is important to avoid the risk of having so many technical standards that, to satisfy them, vendors need to develop a "one-off" solution—in effect, a prototype. Technical standards should set out specific expectations such as those in the following example.

Lubrication standards

1. Devices must be
   a. Protected against damage from falling objects, sources of heat, and the ingress of water
   b. Installed in easily accessible places for ease of maintenance and rapid replacement when needed
   c. Accessible without the need to reach over rotating equipment or shafts
2. Piping must be removable without removing other plant equipment. Joints must be minimized and only used where required for length adjustment or installation.
3. Oil reservoirs must be fitted with a level gauge and positioned so that they can be monitored easily as part of routine operations.
4. Filters and strainers must be easy to replace/clean without stopping the equipment.

3.3.9 Objective Testing

When making decisions about options or features, to reduce the risk of framing and encourage innovation, apply the following steps:

1. Define the function and measures of effectiveness to be used to evaluate options.
2. Generate options.
3. Organize the options into a logical relationships/decision tree.
4. Remove the least attractive options to develop a short list.
5. Select a preferred option from the short list through more detailed evaluation.

Objective testing uses a two-step approach where data is scarce. The first assessment identifies the relative strengths and weaknesses of each option. This discussion also provides a process for decision makers to share insights into the relative merits of each option. That leads to agreement on what evidence needs to be provided to confirm or objectively test that assessment.

Assessment 1 establishes a benchmark for comparison by assessing an existing or known close asset using the six EEM design goals then evaluate each option against this benchmark to subjectively assess the relative strengths and weaknesses of each option.

Assessment 2 is used to collect objective evidence to test the assumptions contained within assessment 1 and confirm/refine those assessments to select the preferred option. In many cases this assessment is sufficient to achieve a consensus on the preferred option (Figure 3.15).

The following is a case study of objective testing using an assessment of filler options for a bottling line.

3.3.9.1 Objective Testing Case Study

Functional description: Liquid fill, cap, and label.

Options for evaluation

1. Benchmark against Line 4 asset (separate labeler and filler).
2. Option: an integrated filler and labeler.
   For: Less space, lower energy costs, more consistent quality, ease of access, and shorter changeovers due to neck handling rather than star wheel bottle manipulation.
   Against: Higher capital cost, higher material waste for each jam.

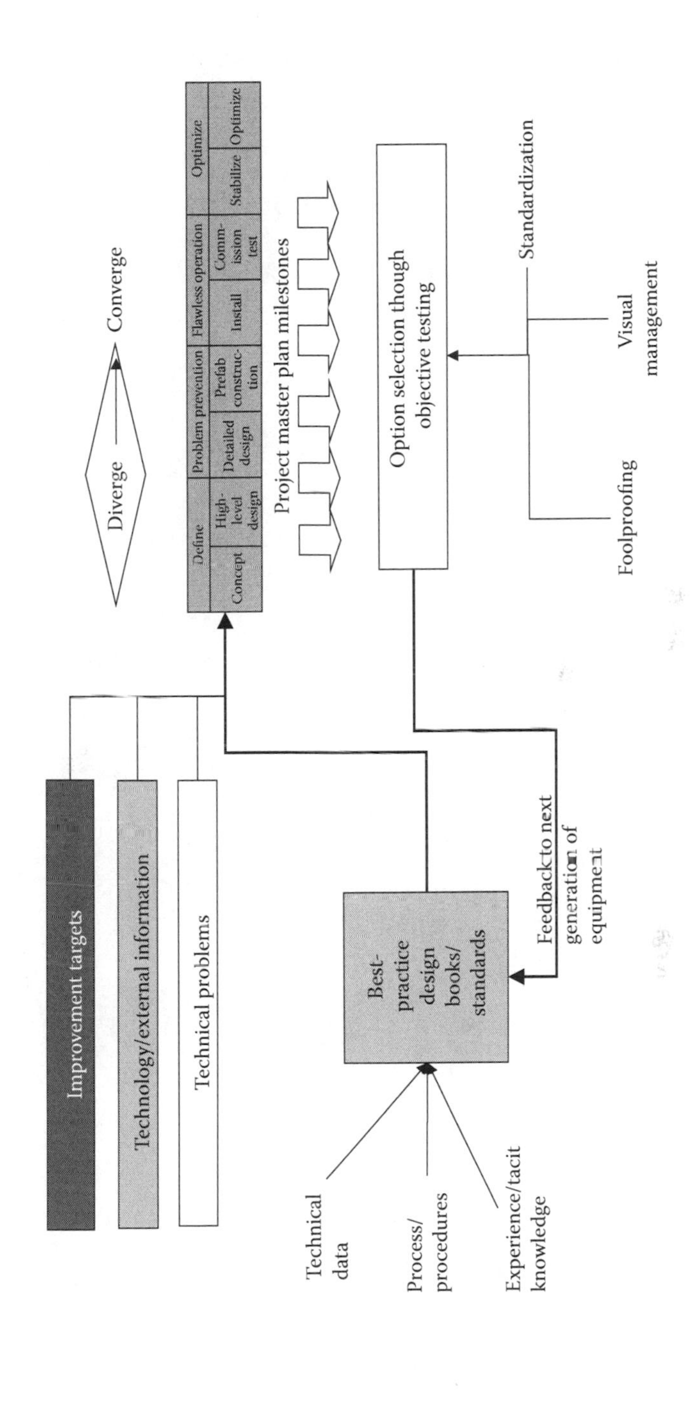

FIGURE 3.15
Objective testing at the heart of EEM.

3. Option: a filler and separate labeler.
   For: Lower capital cost.
   Against: Takes more space. Although the vendors claim it can be run by one operator, it is felt that additional labor will be needed.

In the initial assessment that follows, option 2 gets the highest score and is provisionally the preferred option (Table 3.5).

Option 2 scores higher than option 3 for safety and operability but lower than option 3 in terms of cost. These differences highlight where further work is needed to test the objectivity of this assessment. That leads to the following actions:

- A DILO review to confirm the relative operability and safety scores
- A more detailed assessment of LCCs

The follow-up DILO review confirmed the relative scores for operability and safety but the LCC analysis revealed that rather than costing more, as had been shown by the initial illustrative costs, option 2 had a significantly lower LCC.

When the assessments were adjusted to account for the objective tests (see Table 3.6), option 2 was confirmed as the preferred option.

The final step of the objective testing process is to refine the preferred option where possible so that it scores highest against all six EEM design goals.

In this example, the preferred option scores highest in all categories. If this were not the case, the preferred option would be subjected to further review to determine whether it can be improved further.

The qualitative assessment of gains against the current benchmark can then be converted into quantifiable targets for use as part of the detailed design process and witnessed inspection program (Table 3.7).

3.3.10 Integrating Stakeholder Workstreams

Projects exist in a complicated and dynamic world with many interrelated and in some cases contradictory perspectives. Add to that the pressure of lead times, resource constraints, and economic cycles and it is easy to become overwhelmed by the challenge of planning the plan.

TABLE 3.5

Initial Assessment

Description	Safety	Reliability	Operability	Maintainability	Customer	Cost	Total
1. Current benchmark	3	2	2	3	3	2	15
2. Combined filler labeler	4	3	3.5	3	4	3	20.5
3. Separate filler labeler	3.5	3	2.5	3	4	3.5	19.5
Best score	4	3	3.5	3	4	3.5	

TABLE 3.6

Final Assessment

Description	Safety	Reliability	Operability	Maintainability	Customer	Cost	Total
1. Current benchmark	3	2	2	3	3	2	15
2. Combined filler labeler	4	3	3.5	3	4	3.5	21
3. Separate filler labeler	3.5	3	2.5	3	4	3.0	19.0
Best score	4	3	3.5	3	4	3.5	21

TABLE 3.7

Quantifiable Gains Compared with Current Benchmarks

EEM Goal	Current Benchmark	Expectations/Gains Targeted
Safety	OK, but some problems with access to molds	Improved access due to neck handling (see photographs)
Reliability	Requires frequent attention	20 minutes MTBI
Operability	Complex labeler setup	More automated set up; auto-testing self-clean = 1 less person per shift team, 20 minutes less per changeover
Maintainability	OK	20% fewer moving parts; improved condition monitoring reducing spares holding by 15%
Customer value	Meets current customer needs but cannot handle anticipated levels of promotional demand	More flexible; 50% spare capacity for promotions
LCCs	Increasing running costs	15% lower material costs due to light weighting; 10% less energy use per unit

EEM uses *action mapping*, a causal mapping technique, to make sense of this world. This is used to create a collective plan that takes into account the views and aspirations of multiple stakeholders (Figure 3.16).

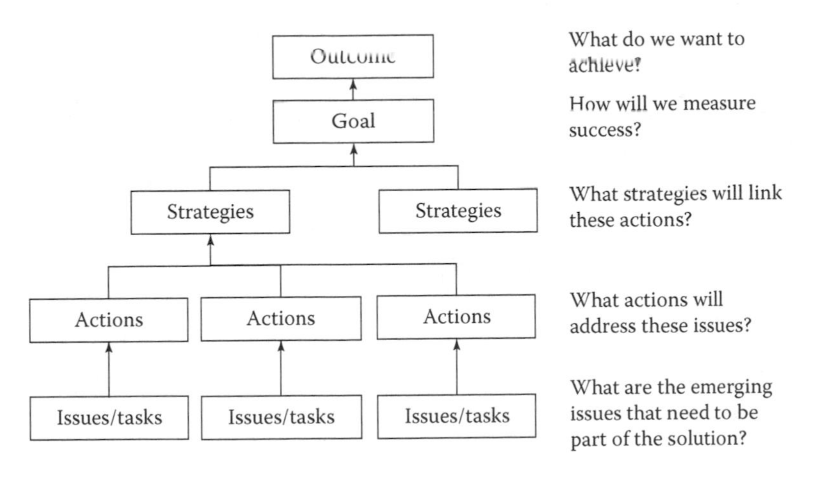

FIGURE 3.16

Action mapping.

The development of an action map is completed by typically 6 to 12 members of the EEM project organization in a working session. This will include commercial, operations, and technology team members.

Prior to the session a clear statement of the desired outcome is drawn up and measurable goals are defined to quantify the achievement of that outcome. The first part of the session involves sharing the current state of commercial, operations, and technology issues related to the defined outcome and goal.

During that briefing, participants record issues and concerns that need to be addressed to deliver the outcome. These issues are shared and discussed to raise understanding and trigger further issues as appropriate. Notes are then clustered under common themes to represent the emerging issues along the bottom of the action map.

A review of these headings is then carried out to identify links between issues and create actions. Finally, linkages between actions are identified and strategic headings agreed to complete the action map (Figure 3.17).

The final part of the mapping activity is to review the action map from top to bottom to confirm that the outcome and measurable goals will be delivered by the strategies and actions identified.

Once confirmed, the participants then work on describing each action point in more detail, including the aims, approach, and milestones to completion. Owners are identified based on that workstream. These completion milestones are then aligned with the EEM stage gates to prioritize actions in line with the project steps.

Figure 3.18 shows the workstream definitions developed from the action map shown in Figure 3.17.

The output from the session is an aligned set of workstreams setting out the scope, interactions, and actions to be carried out.

In addition, the discussion of issues during the session will raise awareness of the concerns and priorities of each of the stakeholders. That means that if, for example, working assumptions have to be revised as a result of one of the workstream activities, the potential impact is understood. Proposed changes can then be communicated to other relevant workstreams.

The workstream milestones are also cross-referenced to the EEM steps so that the stage gate reviews provide a safety net to aid cross-workstream communication.

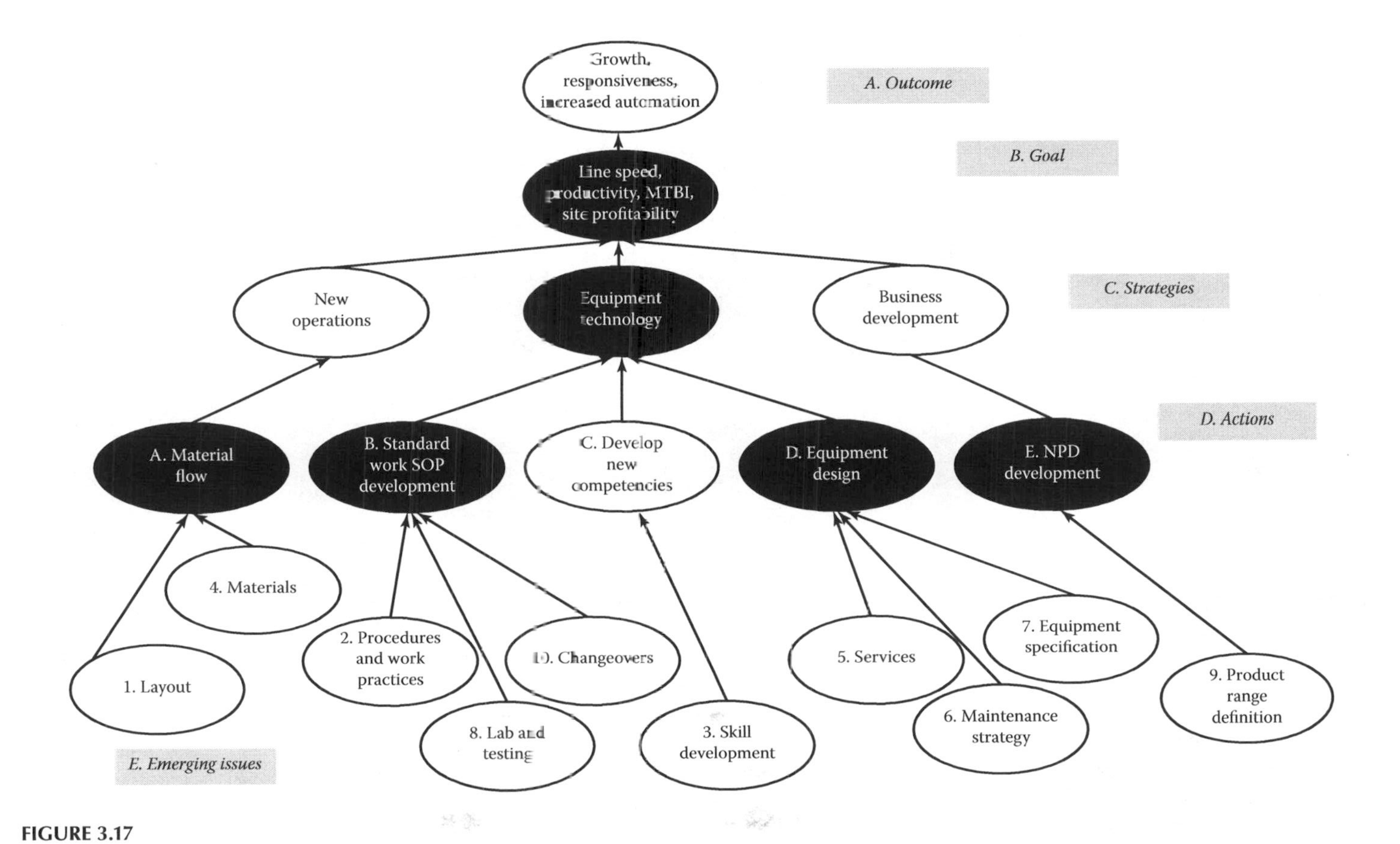

FIGURE 3.17
Action map example.

Required for DD ★ Developed post DD ■

Workstream	Aim	Planned outputs				
		1	2	3	4	5
1. Material flow	Confirm line design to ensure suitable material storage, people, and material flow.	Adequate storage in area and external year 1, year 2, etc. ★	Agree people, material flow tested and agreed.	Short interval control process in place and working supporting material, ■	Automated pull lists in place and working.	
2. Working method development and standardization	To establish operational and QA SOPs, standard changeovers, and review current QC procedures.	Formalized ways of working and accountabilities. ★	Plan for improvements to current methods, including capex justification. ■	Retraining plan implemented.	Effective communications/ integration of end-to-end value stream.	
3. Skills development	To develop knowledge and skills to be able to operate and manage the new line and deliver project targets and anticipated business deliverables.	Define skills required.	Identify skill gaps.	Approve training plan and release resources. ■	Deliver training.	Confirm competency.
4. Equipment design	Ensure that factory systems specification captures requirements for services, equipment intervention spec, and maintenance strategy.	Defined spec for optimized line performance, fit for purpose automated PM regime. ★	Defined asset care regime. ■	Submit capital equipment requirement for services.	Define automated/ minimum changeover protocol defined.	Optimized line/ area/layout.
5. NPD development	Understand current and future plans that could impact on design formats, including materials, product specification, and the direction of incremental product improvement.	Structured wish list of areas. ★	Agreed in/out list and incremental capex needs. ■	Equipment spec/defined areas needing flexibility and commissioning protocols.	Year 1–2 new product development plan setting out decision trigger points and commissioning tests.	

FIGURE 3.18
Workstream definition.

3.4 SYSTEMATIC DESIGN DEVELOPMENT: DEFINE PHASE

The design and performance management tasks during the concept and high-level design steps are as follows:

- Confirm the project brief.
    - Review the operational data to confirm the understanding of current operation and OEE losses.
    - Define the future project vision, targets, and tactics to deliver the desired outcome.
- Evaluate options and select a preferred approach to delivering the concept.
    - Set targets/measures of effectiveness.
    - Generate options.
    - Develop a long list of options for review and identify a short list for further investigation.
    - Evaluate the short list of options; refine the preferred option; create an outline of equipment; define civil engineering, utilities, and interfaces with current systems and processes.
- Collate a list of requirements.
    - Refine the layout and explore the impact on current operations.
    - Define design modules and targets.
    - Include the *factory acceptance test* (FAT), *site acceptance test* (SAT), *witnessed inspection*, and Construction Design and Management (CDM) Regulations.

A common pitfall during this phase is the way the scope is set during the investment-planning process.

For example, a food and drink company at the start of its EEM journey decided to invest in upgrading equipment to bring in-house a product that currently was packed by an outside company. A recent price hike from the current contract packing company had reached the attention of the senior management team. They decided to sanction investment to upgrade an existing line so that the product could be packed in-house. Normally, such an investment would have been carried out without question, but as part of a newly introduced EEM process, the core team identified that this was not the best use of the investment. As part of the PP data collection process they identified that if the funds allocated were spent elsewhere it would produce greater benefits with a faster payback.

Based on this experience, later investment project briefs were then set out in terms of the desired outcome and prioritized targets. This also established the EEM core team role as part of the investment management process, which included activities to

- Translate desired outcomes into metrics
- Define problems to overcome to deliver those targets
- Set out scenarios to consider
- Use those scenarios to develop an equipment master plan for the site linked to strategic measures/action triggers

EEM tools to support this activity include

- DILO
    - Identifying what is needed and the impact on operations
    - Clarifying EEM design goals, guidelines, and technical standards
- Objective testing
    - Systematically assessing and selecting options
    - Using design guidelines to encourage creativity and innovation
- Design module review of preferred option
    - Highlighting potential areas of weakness and opportunity
    - Understanding equipment losses and countermeasures and becoming aware of their impact on LCCs
- Action mapping
    - Defining equipment performance drivers

The equipment master plan was routinely updated as part of the strategic review process. The outcome was that investment decisions were made within the context of the master plan rather than as a knee-jerk reaction to circumstances.

3.4.1 Define Phase Audit Questions

At the concept stage gate, use the following questions to confirm progress. Have we identified a concept capable of meeting

- The equipment master plan's desired outcomes and the return on investment goals
- The EEM design goals
- Buy-in from commercial, operations, and technology functions

At the high-level design stage gate, use the following questions to confirm progress. Have we refined the concept into a preferred option that has the potential to achieve

- The lowest LCC (and flawless operation from day one)
- Flawless operation from production day one, taking into account the project critical path and knowledge gaps
- Buy-in from commercial, operations, and technology functions.

3.5 SYSTEMATIC DESIGN DEVELOPMENT: DESIGN PHASE

The first step following funding approval, the selection of a vendor(s), is covered in more detail at Chapter 5, "Project and Risk Management."

The design and performance management priorities for this phase are to tease out latent design weaknesses and enhance project value. This involves

1. Vendor induction: Agreeing the project vision and route to flawless operation, establishing a foundation for collaboration
2. Detailed design workshops
3. Operations change management

3.5.1 Vendor Induction

Formally inducting the vendor design team into the project provides the opportunity to assure that the new team members fully understand the EEM design goals, guidelines, and standards.

Smaller projects should include an induction/mobilization process for vendors as part of the first module of the detailed design workshop.

Larger projects will benefit from a formal EEM visioning workshop over one or two days to

- Meet and greet
- Lay the foundations for a proactive working relationship
- Reinforce the design goals, guidelines, and standards agreed with the vendor (holding them to account)

The desired outcome from the vendor induction session is to

- Communicate expectations and the EEM approach to the vendor project team so that the vendor can make plans to meet those expectations

- Reinforce the desire to collaborate on teasing out latent design weaknesses and opportunities to add value to the project; also that the scope of this includes equipment design, installation/commissioning plans, skill development processes, risk assessment, and witnessed inspection protocols
- Agree the timing of the detailed design workshops for each design module

3.5.2 Detailed Design Workshops

EEM module detailed design involves two working sessions.

- Meeting 1 involves the assessment of an outline design using 10% outline design drawings to explain the EEM goals, guidelines, and standards, to make the operational realities clear, and to outline ways of working. This will bring detailed design priorities to the surface. The vendor then uses that information to produce 90% drawings, which explain how the design goals are to be met.
- Meeting 2 involves the vendor explaining in detail how the more detailed design meets those goals and the forward program to deliver flawless operation from day one (Figure 3.19).

3.5.3 Problem Prevention

During the EEM detailed design process, the role of the vendor is to assure technical stability and the achievement of process control requirements. The EEM core team role is to understand how to achieve process resilience and process control with the new asset.

To achieve this the core team should view the detailed design workshops as a voyage of discovery. They also need to recognize that an analysis of over 500 years of run time in a range of industry sectors reveals that only 15%–20% of outages are due to component design weaknesses. Other factors that have a significant impact on problems include

- Asset care plans
- Operating methods
- The ingress of dust and dirt

These contributors can be targeted during the detailed design process using design guidelines, standards, and the checklists developed during the high-level design module reviews. Further discussion of these areas is included in Chapter 4.

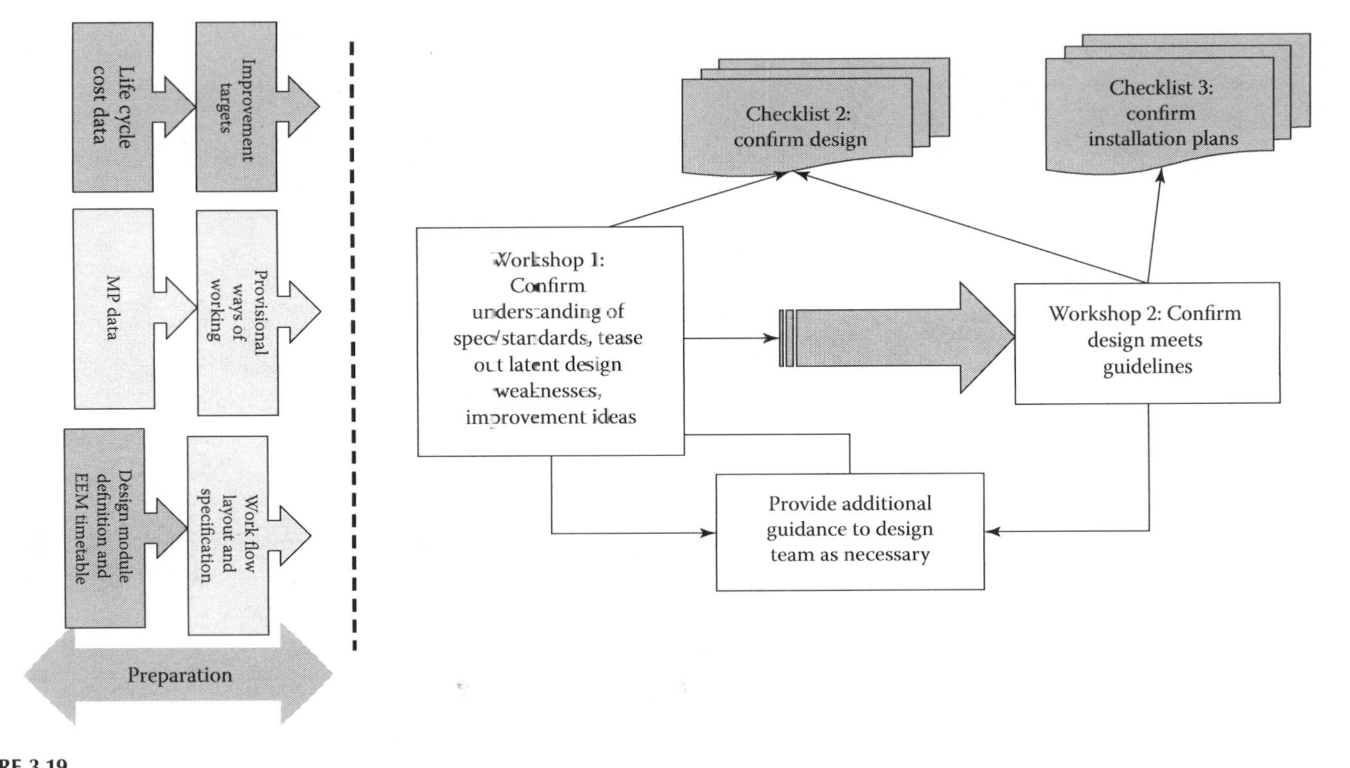

FIGURE 3.19
Detailed design process.

3.5.4 Detailed Activity Planning

Update the action map to incorporate the lessons learned during the detailed design process and integrate the vendor's workstream milestones within it. Define witnessed inspection protocols for each forward milestone after the detailed design is frozen. Use this to provide a basis for coordinating the progress of parallel workstreams.

This should include testing to confirm progress of the steps to develop new organizational competencies. A common mistake at this point is to rely on vendor training alone. The design and delivery of the skill development process should include input from the vendor but best results are achieved when the process is handled as an internally managed process, as set out in Figure 3.20.

3.5.5 Design Step Audit Questions

At the Detailed Design stage gate, use the following questions to confirm progress. Have we achieved:

- Effective communication between vendors and internal personnel to underpin collaboration
- A design that meets design goals, guidelines, and technical standards
- A realistic and achievable plan taking into account the project's critical path and knowledge gaps
- Buy-in from all parties

Risk assessment/mitigation and witnessed inspection processes to assure the achievement of design features and performance is covered in Chapter 4.

3.6 SYSTEMATIC DESIGN DEVELOPMENT: REFINE AND IMPROVE

As the new workplace takes shape, the full three-dimensional potential of the operation is revealed. The use of simple models at earlier steps can help the team to visualize the layout and to make decisions about how to make best use of space, but there will always be fine-tuning needed. Make use

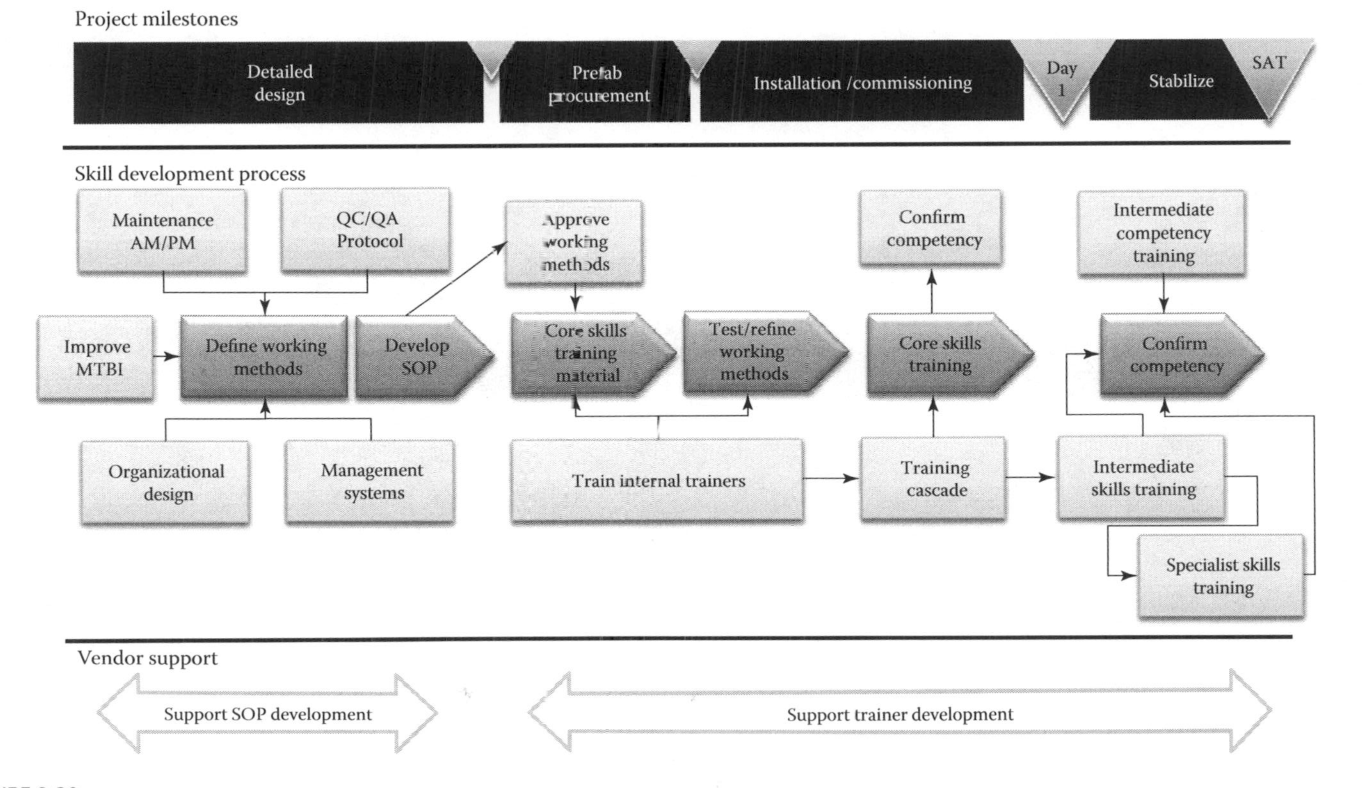

FIGURE 3.20
Skill development process design.

of this as part of the process of handing over ownership of the territory to those who will work in it.

The main design and performance management tasks during installation and commissioning activities involve

- The application of tests/checks to confirm the delivery of design features and performance levels
- Micro-level workplace layouts applying problem prevention design techniques to provisional ways of working so that they are easy to do right, difficult to do wrong, and simple to learn
- Capturing the lessons learned and updating the best-practice design books
- Mobilizing activities to support the development of standard ways of working, visual indicators, and accountabilities to
    - Minimize human error
    - Simplify skill development
    - Support defect prevention
    - Enhance operational flexibility
- Confirmation of process capability and design goals
- Managing the glide path to flawless operation from day one
- The development of process optimization tactics

Design and performance management is supported by witnessed inspection activities (see Chapter 5, "Project and Risk Management") to provide feedback and refine design guidelines and technical standards. Figure 3.21 shows an example of how an automotive company used design feedback to improve the design of their project delivery training plans, reducing skill development time and plant commissioning timescales.

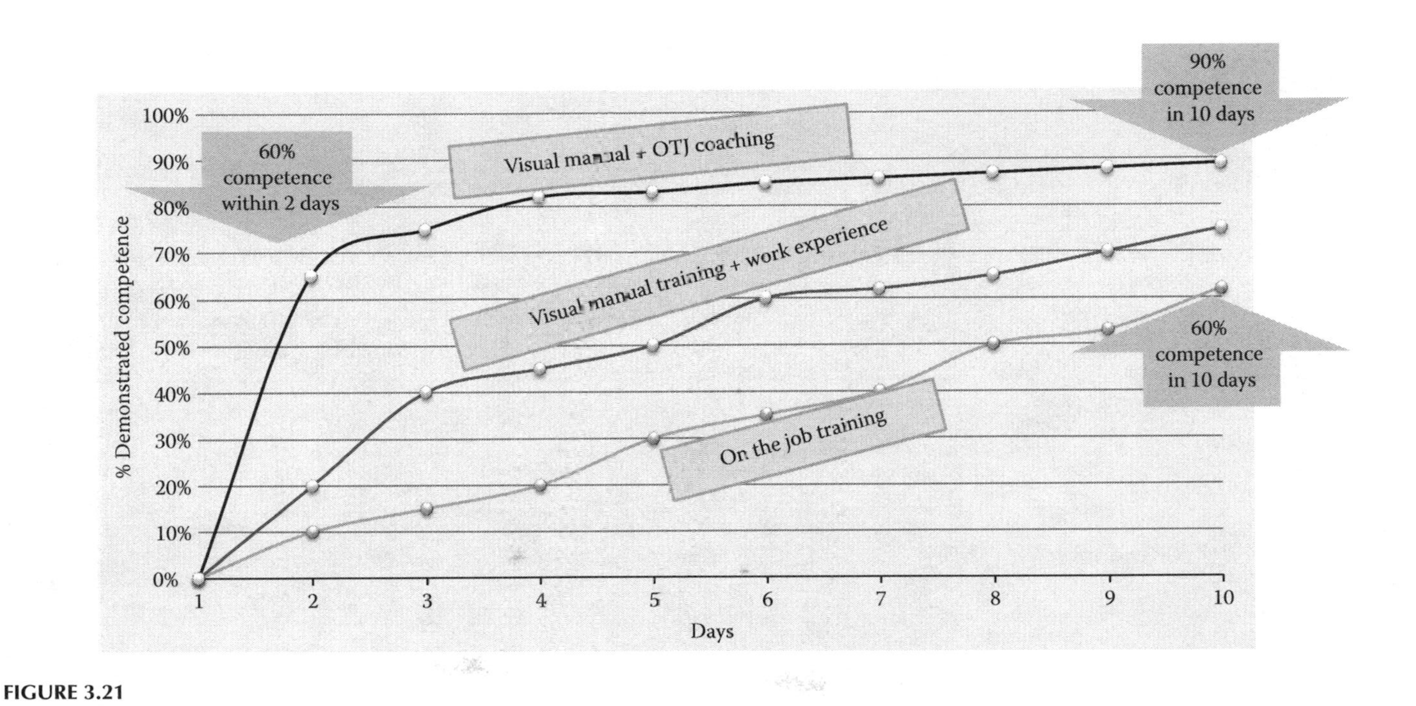

FIGURE 3.21
Reducing skill development time (assembly line operator study).

3.7 CHAPTER SUMMARY

This chapter provides an explanation of EEM design and performance management, the first of the five EEM subsystems, covering

- How developing good design depends on creating the conditions for innovation and processes to guide systematic design development
- How both innovation and systematic design development benefit from an understanding of
  - Current shop floor reality
  - Translating tacit into explicit knowledge
- The use of PP data lists as an aid to delivering better designs that achieve better performance
- How to manage parallel working and integrate the workflows of multiple stakeholders
- Vendor skill sets and how to get the best from working with them to tease out design weaknesses and prevent problems
- The design of activities to develop in-house capabilities and manage project-related operational changes

Key messages include the following:

- Design is a complex wicked problem that benefits from a stepwise approach that considers multiple options as a way of developing a preferred one.
- Be clear about what you want and manage the design requirements of all stakeholders against a common outcome and workstream milestones.
  - Develop design guidelines to support the EEM goals.
  - Create technical standards carefully so that they aid standardization and guide detailed equipment manufacture without creating a unique design proposition that excludes the use of vendor experience and innovation.
  - Breaking the design brief into modules provides a structure to support the planning, organization, and control of the design and delivery process.
  - Carry out an operational review for each module to highlight potential areas of weakness and opportunity.
  - Use this to define the performance drivers for each module.

- Add detail to and refine the design of each module, including operational ways of working through practical activities at each project step.
- Extend the design process to include performance delivery to confirm design assumptions and enhance internal design capabilities.

4

Specification and LCC Management

This, the second of the five EEM subsystems introduced in Chapter 2, covers the timely capture and formatting of information at each step to provide reference information, aid communication, and support the delivery of project goals.

There are close links between the topics here and those in Chapter 3, "Design and Performance Management." As illustrated by the document map in Figure 4.1, there are also links between this chapter and Chapter 5, "Project and Risk Management."

The main focus for this chapter is on the evolution of documents that directly impact on operating methods and life cycle costs (LCCs). The chapter begins with an introduction to specification and LCC management best practice before covering specification-related activities during the project define, design, and refine phases introduced in Chapter 2, "The EEM Road Map."

4.1 HAVING THE RIGHT CONVERSATIONS

As mentioned previously, capital projects are characterized by social interaction and iterative decision loops. Within this context, the purpose of the specification is to record the result of conversations by stakeholders to

- Get the right design (*define*)
    - Capture and develop ideas
    - Make choices, tease out latent design weaknesses, and reduce LCCs

FIGURE 4.1
EEM document map.

- Get the design right (*design*)
  - Define equipment functions and design modules
  - Generate options and evaluate using EEM goals and guidelines
  - Select and refine the preferred option to enhance project value
- Deliver the specification (*refine*)
  - Communicate effectively and manage risks
  - Develop the skills, competence, and confidence to perform
- Ratchet up performance (*improve*)
  - Learn how to create optimum conditions
  - Incrementally improve the product and customer service

4.1.1 Equipment Master Plans

In mature EEM organizations, the project specification builds on the content of a site's *equipment management master plan*, which summarizes future commercial, operational, and technological challenges and the impact on these of equipment investment plans (Figure 4.2).

The availability of an equipment management master plan makes it easier to position each project within the context of the longer-term strategic plan. It is also an effective antidote to knee-jerk investment decision-making.

4.1.2 Specification Formats

In addition to the content of the specification, the specification format is also important. This evolves during each step (see Figure 4.3). For example:

- The use models and sketches at the early stages encourage creativity and innovation (see Figure 4.4).
- Milestone plans set out key project signposts and timescales so that the project can be more easily aligned with other related projects and business time lines.
- Models, simulations, and trials support the testing of ideas in the real world to improve understanding and develop insight.
- EEM goals and guidelines help to assess vendor outlook and suitability.
- Simple estimates provide the basis for outline project justifications.
- Graphics and pictures speed up the process of stimulating and sharing ideas, encourage input from others, communicate lessons learned, and accelerate skill development.

	How is this changing	Challenges	Equipment management goal
Commercial • Customer • Expectations • New products and services	Increasing expectations of quality cost and delivery performance	Incremental product development, new products and services for growth Supply chain for global operations	Capacity for future demand Robust supply chain Simple logistics/forecasting needs Flexible to potential market shifts
• LCC Management • Main contributors (4M) • Technology	Material costs, energy costs, labor costs, batch sizes	Sustaining volume on existing products while introducing new ones Energy inflation	High level of resource recycling Flexible to financial risks (e.g., vendor) Easily scalable to 400% or to 25% Access to high added value markets
Operations • Operability • OEE losses • Lead time/flow/flexibility	Focused improvement, smaller batches, new product life cycles, shorter lead times	Systematic hidden loss and lead time reduction Achieve "Normal operations" with less effort	One touch operation for height, position, number, color, etc. Flexible to volume risk Flexible to labor skill levels
• Maintainability • Predictable component life • Defect prevention	Reduced plant availability for maintenance, more complexity, increased precision, availability of skills	Extending MTBF/MTTR Maintenance prevention Supporting process optimization Lack of skills	Inbuilt problem diagnostic Self correcting/auto adjust Predictable component life Easily overhauled, fit and forget
Technical • Reliability • Increase precision • Automation • New technology/materials	Improved control, lower grade materials	Capturing lessons learned Adapting plant to new products, improving technical competence	High MTBI Stable machine cycle time Easy to measure Flexible to material variability
• Safety • Ease of compliance • New legislation	Increased legislation and controls, higher environmental expectations	Safe, environmental practices with minimum impact on performance Improved sustainability	Foolproof/failsafe operation High level of resource recycling Uses sustainable resources

FIGURE 4.2
EEM master plan challenges.

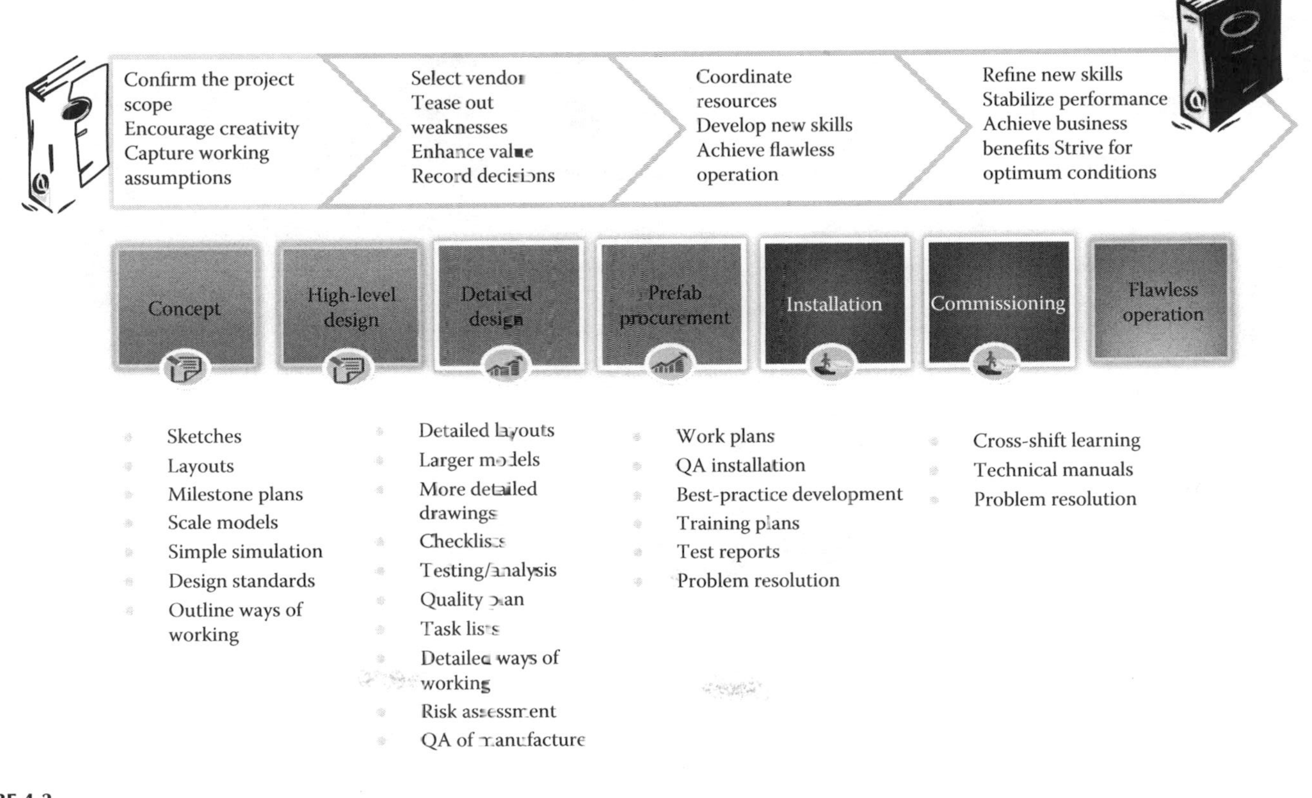

FIGURE 4.3
Specification formats by step.

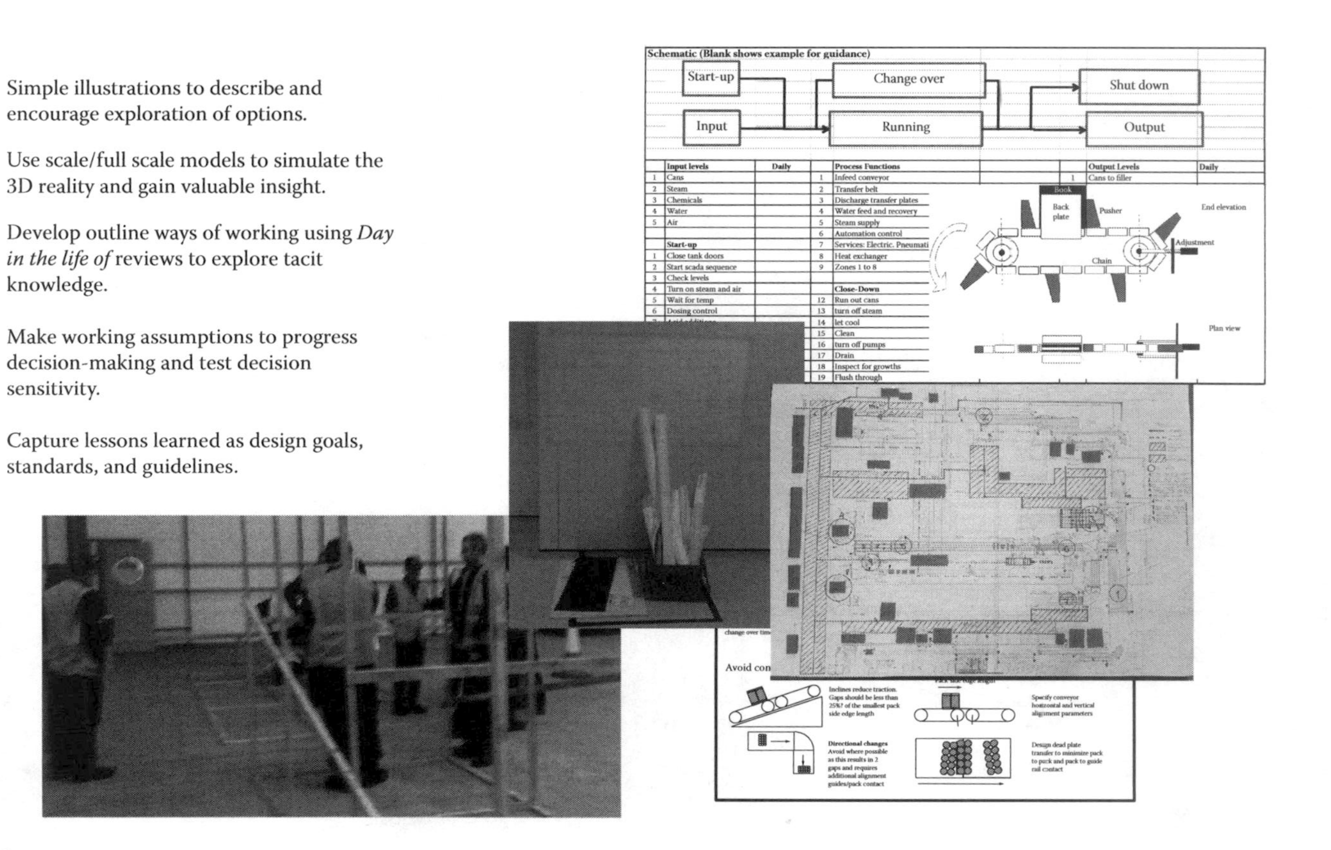

FIGURE 4.4
Sketches, diagrams, and models.

- Analysis charts (X-charts, house-of-quality charts) record the insights gained on the design journey.
- Draft and version-controlled documents support the management of organizational change and the development of operational competencies.
- Formal specifications provide clarity for commercial contracts, risk assessments, project plans, witnessed inspections, and operational documentation at later steps.

The value of taking time to make simple models out of low-cost materials is often underestimated. The process of developing even simple models helps the team to gain an insight into the three-dimensional footprint, access, workspace, and material flow simplification. Once built they provide a simple way to communicate changes and stimulate new thinking. Such models are at least as valuable as the more expensive computer simulation models.

4.1.3 Keep the Conversation Flowing

It is not unusual for knowledge or information gaps to come to light when making decisions at the early project steps. This can constrain the conversation because those involved feel that they cannot make a decision if there is uncertainty about related decisions.

For example, a team developing the specification for a new food line were uncertain if own-label customers would want existing products validated on the new line. The potential continuous flow technology would be different to that which was currently being used. This added complexity made some of the team anxious that their decision could result in problems later on.

To progress further, the team agreed a working assumption that customers would not want to revalidate their existing products. With the understanding that this working assumption would be revisited later, they were then able to progress with the specification design. Having defined their forward plan and timetable, they were able to agree with senior management that the working assumption was acceptable. Also, as they worked through the process, they gained a better understanding of the drivers for product quality and yield. With the insight gained, they became confident that they could match the current product specification and possibly at lower cost. The use of a working assumption helped the team to make progress and gain a greater insight into the implications of that assumption.

The causes of such blockages are many and various, including lack of experience, team dynamics, confidence, or unrealistic expectations about what data can be provided. Working assumptions provide a bridge for teams to bypass areas of uncertainty and gain sufficient insight to confirm or refine those assumptions.

With hindsight, those making the decision can appear to have over reacted to the uncertainty because because the concerns raised proved to be unfounded. Regardless, uncertainty is part of the iterative reality of project decision making and declaring a working assumption is a simple but effective way of exploring options when the way ahead is unclear.

Capture working assumptions and key decisions as they are taken so that

- As people join the project, the reasoning behind key decisions can be explained easily.
- The decision basis can be revisited if the outcome was not as planned or could be improved.
- Future project teams can understand the nature of the decisions taken and how they impact on the final solution.
- When considering a change to an agreed working assumption, the impact on all work streams can be easily revisited.

4.1.4 Setting the Design Agenda

As with any value-added conversation, the scope of topics discussed should be defined by a clear agenda. Setting the design agenda is critical. Some may want to cover personal agendas, vendor sales personnel may want to discuss the features and benefits of their offerings. To deliver best value, the conversation needs to focus on how to minimize operational LCCs.

Within EEM, the design agenda is set by the contents of the best-practice design books (see Chapter 6, "Project Governance"). These are used to capture the lessons learned from focused improvement activities before the project and through the collation of *problem prevention* (PP) data as part of the project.

Best-practice design books provide a library of

- Design module features, including process, methods, and checklist points to support the achievement of
  - Flawless operation from production day one
  - Optimum performance

- Design goals, guidelines, and standards to
  - Provide the context for the development of concept options that take into account known concerns and opportunities from previous projects
  - Help potential vendors rapidly gain an insight into our requirements
  - Support the detailed design process
- Site equipment master plan policy guidelines to align
  - Site and corporate equipment procurement standards
  - Site manufacturing process and product development strategies

As an organization develops proficiency in the application of EEM, the content of its best-practice design books become more concise and better understood by potential vendors. The design agenda is then able to cover more topics in less time, increasing the speed to market and the rate of return on investment (also covered in Chapter 6).

4.1.5 Checklist Management

A common pitfall is to assume that the development of extensive checklists will result in improved project delivery performance. It won't.

A good checklist is precise, efficient, and easy to use. If it is too long, it is easy to duplicate points. It should provide reminders of only the most important steps, rather than trying to spell out everything. A checklist can't do your job for you.

Above all, a checklist should be practical. Generally, a checklist is best suited to work that's carried out in a predictable order. It should help the user to follow a logical sequence. Ad hoc checklist points are confusing and can fail to communicate their true meaning or importance.

The best way to achieve this is to incorporate checklist points as part of a

- Procedure or briefing note—for example, preparation, sequence steps, and outputs
- Quality assurance (QA) test
- Stage gate review

In line with these principles,

- EEM goals, guidelines, and standards act as a checklist.
- EEM briefing notes with the aim, approach, and output formats are also a form of checklist.
- Checklists are targeted at individual EEM steps.

Checklist points are generated during the first three steps of the EEM road map as a result of the following:

- *Day in the life of* (DILO) reviews are used to update/refine design guidelines.
- Module review checklist points are used as part of the detailed design process to support the definition of witnessed inspection tests.
- Risk assessment outputs and quality plan exit criteria are incorporated into the witnessed inspection process to track the progress of the glide path to flawless operation.

4.2 CREATING POWERFUL SPECIFICATIONS

The capture of increasing detail at each step utilizes a hierarchy of documents such as that depicted in Figure 4.5.

4.2.1 Floor Layout Drawings

In the words of the song, a picture paints a thousand words. The same is true of layout drawings. Creating a simple layout drawing helps people to visualize what the new option will be like to work in. Use the development of this to build ownership to the new operation.

As a minimum, layouts drawings should include

- Space boundaries and links to other processes
    - Storage and production zones
    - Fire exits and safety equipment
    - Clean areas/wash stations
- Equipment footprint
    - Access hot spots
    - To assure the line of sight during operation
    - Servicing points
- Conveyor routes and access points
    - Floor level
    - High level
    - Crossover points

Project specification file	Define	Design	Improve	Stabilize/ Optimize
7. Layouts: Location, material, and work flows	Outline layout and material flows	Detailed layout, material flows, work area definition	Work station design, visual management/5S standards	Standards compliance and layout optimization
8. Process description: Functional description, module descriptions, supporting documentation including FAT/SAT	Module review notes: Column 1	Technical specification	Technical manuals	Technical library
9. Method statement: Operator, maintainer, workplace layout, procedures, and skill development plan	Module review notes: Columns 2 and 3	SOP design, core and intermediate skills definition	Training material, training cascade, and competency assessment	Skills transfer, best practice manual update
10. Resource management: Material control, QA, LCC	Business case	Organization and supply chain design	Team development and supply chain mobilization	Workflow optimization

FIGURE 4.5
Specification hierarchy.

- Workplace footprints
    - Safety hot spots, guard opening, etc.
    - Main aisles
    - Walkways
    - Material flow
    - Drop zones

Even though the layout may be a simple sketch such as the one in the example, once defined, the layout drawing should be subject to design freeze and version/change control (Figure 4.6).

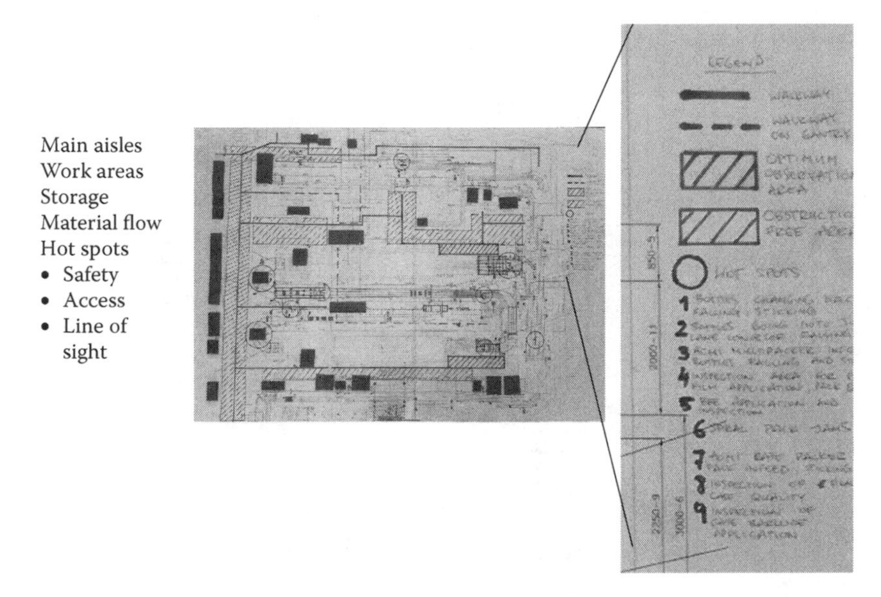

FIGURE 4.6
Layout and material flow.

4.2.2 Module Review

The purpose of the module review is to collate information and refine design module specifications prior to vendor selection and detailed design. This includes

- The capture of issues to address and common problems that can impact on the delivery of project goals
- The clarification of design guidelines and acceptable/optimum design criteria

- The value engineering of module features to enhance project value
- The identification of a best-practice development list for new practices and current practices in need of improvement

The specification for each design module consists of

- Process information
    - Inputs, process steps, and outputs
- Working methods
    - Start-up, steady state, close-down/clean-out
- An asset care plan
- Prevention, inspection, and servicing routines

The two-step module review process is set out in Figure 4.7.

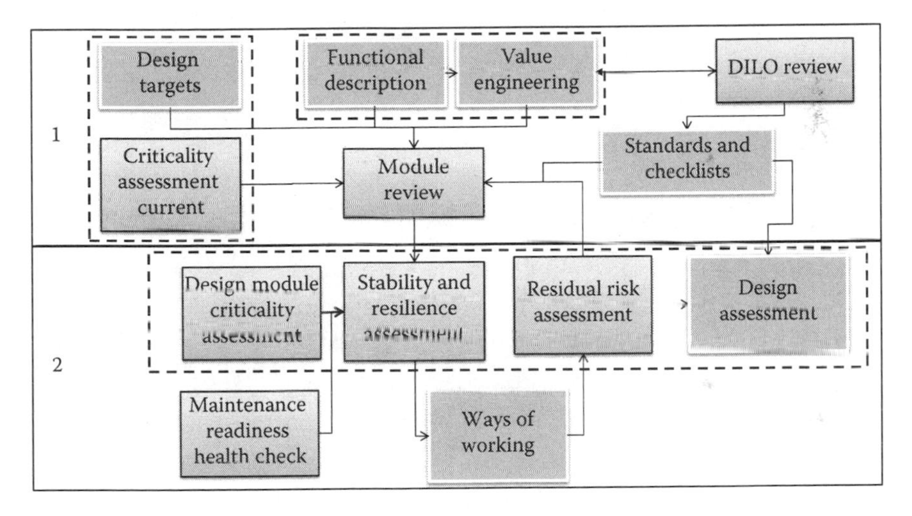

FIGURE 4.7
Module review process.

4.2.3 Module Review Step 1

The first step is carried out prior to the creation of the *invitation to tender* (ITT) document used to support vendor selection. This activity is an opportunity to involve shop floor representatives, capture tacit knowledge, and engage front-line personnel with the new operation. Managed correctly, this will create interest and communicate the scope of future changes to those who work in the area.

The module review headings provide an agenda to trigger discussion on issues to be overcome to achieve the goal of flawless operation. Allow around two hours per module at the high-level design (HLD) stage (Figure 4.8). Ask participants to feedback ideas and issues they think of later.

Following Step 1 of the module review process, collate the outputs with the relevant points from the DILO review to set out

- Issues/concerns to be addressed
  - Are they part of or outside of the project scope?
  - Do we need to know more?
- Common problems and areas with potential for improvement
- Features to be part of the specification
- Areas to check at the detailed design, equipment manufacturing, installation, and commissioning steps
- Potential performance gains based on equipment history, such as OEE data
- Risk mitigation tactics from the risk assessment process described in Chapter 5

4.2.3.1 Value Engineering Review

During the review, value engineering principles are applied to assess the value added by each design element. The purpose is to understand the value of each design feature/element and where possible reduce the cost of providing that value. An example of the questions used at the HLD, detailed design, and installation planning stages is set out in Figure 4.9.

4.2.3.2 Preparation

Develop a functional description for each design module covering:.

- What function does the module perform?
- What are the inputs, processes, and outputs of the design module?
  - What is the value provided by each?
  - Consider using a criticality assessment to understand the relative criticality of components to meeting the six EEM design goals.

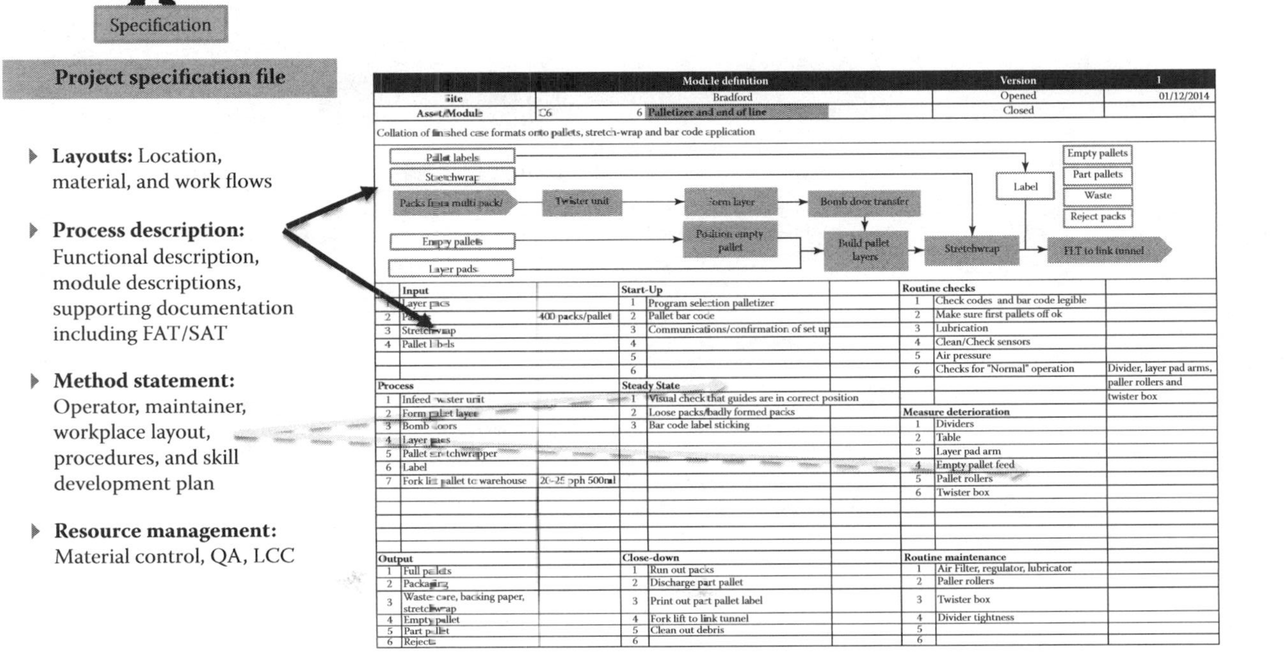

FIGURE 4.8
Module review agenda.

	Value engineering area	1. High-level design	2. Detailed design	3. Installation
1	Understanding of hidden losses, nonvalue-adding and wasted efforts	How can we design this out?	How to improve this to prevent potential problems?	Have we designed good practice in?
2	Understanding of working methods, risk, and how to build in good practices	What outline ways of working could be relevant?	Does the design make it easy to do right and difficult to do wrong, are skill gaps identified?	Has a suitable training and skill development regime been introduced?
3	Assess if project will impact on other processes	How will this need to change as a result of this project?	How to align design with current systems and processes?	Have we updated our systems and procedures to account for this?
4	Understand critical parameters so that we can control them	Do we know what is critical and how to control it?	How to improve this to prevent potential problems?	Have we updated our systems and procedures to account for this?
5	Identify where knowledge may be available elsewhere	Have we taken into account lessons learned and best practices?	How to assure compliance with best practice?	Have we plans to update systems and procedures to account for this?
6	Identify maintenance methods and risk, and define how to achieve good maintainability	What outline prevention, inspection, servicing methods could apply?	Does the design support maintenance prevention standards are spares and skill gaps identified?	Are working methods defined and trained in? Has spares provision been made? Is CMMS updated?
7	Understand the specification of the new process	Do we know how this compares with what we need? Can we reduce the impact of any gaps?	How to improve this to prevent potential problems?	Do we have plans to update systems and procedures to account for this?
8	Understand operational methods, risks and how to build in operability	What outline operational/clean out activities could be relevant?	Does the design make it easy to do right and difficult to do wrong? Are skill gaps identified?	Are working methods defined and trained in? Have production materials provision been made? Are systems up dated?
9	Understand the main drivers of life cycle costs potential savings	Do we know what is required? Can we improve it?	How to improve this to prevent potential problems?	Do we have plans to update systems and procedures to account for this?
10	Identify current risks can be reduced with the new process	Has a relevant risk assessment been carried out and actions been identified?	How to improve this to prevent potential problems?	Have we updated our systems and procedures to account for this?
11	Identify complex processes which could result in errors or additional costs	Does the project increase risk levels? How can we simplify/improve control?	How to improve this to prevent potential problems?	Have we updated our systems and procedures to account for this?
12	Identify where internal and external standards apply	What standards apply?	How to assure compliance with relevant standards?	Have we designed good practice in?

FIGURE 4.9
Value-engineering topics.

- Define start-up, steady-state, and close-down method statements for each module.
- What asset care tasks are required?

Collate operational data, including design targets, criticality assessments, and design guidelines/checklists from the DILO review.

4.2.3.3 Approach

Use the question list to highlight areas with potential for improvement and actions to confirm/refine the approach used.

In addition to improving the design brief, the module review generates checklist points for further discussion or to support witnessed inspection activities. Examples of typical checklist points are included in Figure 4.10.

Note how each checklist point indicates the relevant design guideline and at which EEM step it is applied to, plus a question to guide the application of the checklist point.

4.2.4 Module Review Step 2

The second step of the module review is carried out after vendor selection as preparation for the detailed design process to

- Improve understanding of the preferred vendor's asset/process
- Assess where there are gaps in knowledge or capability, or where current operating practices need to be improved

Working with vendors is discussed in more detail in Section 4.2.9 onward. The outputs from the module review process are summarized in Figure 4.11.

4.2.5 Resource Management/Systems Specification

Most capital projects involve changes to business processes and/or operations management. This section covers how to release the full potential from these changes.

Table 4.1 sets out the headings of a user specification review to clarify

- The system's purpose
- How the system will be used
- Changes to roles and responsibilities
- Training and implementation tasks

Module-specific standards/checklist points					
Site	Case study		Opened	13/10/2014	
Asset	Palletizer		Closed		
Code	**Heading**	**Module-specific issue**	**EEM Step**	**Question**	**Result**
Cust 2	Verification	14-digit GTIN/EAN code or Princes product code using special application identifier. Quantity, batch, and best-before date. Logistic unit label with a unique SSCC number.	3. DD	Will this be provided?	
Cust 7	Defect detection	Packs missing.	3. DD	How will this be provided?	
Cust 7	Defect detection	Pack bar code missing.	3. DD	How will this be provided?	
LCC 10	Electricity	All conveyor drives to be zoned with a timed automatic shut-down function.	3. DD	Will this be provided?	
Mnt 1	Access, egress, and lighting	Full pallet conveyors and machine interfaces designed for easy access for cleaning and maintenance.	3. DD	How will this be provided?	
Mnt 1	Access, egress, and lighting	Access for cleaning.	3. DD	How will this be provided?	
Mnt 6	Inspection	Ease of checking lubrication.	3. DD	How will this be provided?	
Mnt 6	Inspection	Ease of pack-turner belt checks (can require regular attention).	3. DD	How will this be provided?	
Mnt 7	Prevention	Sensor cable protection.	3. DD	How will this be provided?	
Mnt 7	Prevention	Avoidance of jams causing consequential damage.	3. DD	How will this be provided?	
Mnt 8	Repair	Ease of sensor replacement.	3. DD	How will this be provided?	
Mnt 9	Routine servicing	Ease of greasing.	3. DD	How will this be provided?	
Mnt 10	Cleaning	Prevent need to clear debris under equipment or provide cleaning aid.	3. DD	How will this be prevented?	
Op 1	Access, egress, and lighting	Plastic roll handling and lift equipment.	3. DD	Will this be provided?	
Op 1	Access, egress, and lighting	Access to release jammed hooder.	3. DD	How will this be provided?	
Op 1	Access, egress, and lighting	Access to release jammed pallets (clear routes).	3. DD	How will this be provided?	
Op 3	Changeover	Changes in pallet layers/layer format; format changeovers should be automatic via the HMI panel and should require no mechanical intervention.	3. DD	Will this be provided?	
Op 5	Operator intervention	When palletizer gets out of sequence, a single-button reset facility to enable is to be included.	3. DD	Will this be provided?	
Op 5	Operator intervention	All pallet conveying should be designed to be operated in a manual mode.	3. DD	Will this be provided?	
Op 5	Operator intervention	The pallet conveyors are required to have an inverter control, and be reversible.	3. DD	Will this be provided?	
Op 6	Visibility of normal conditions	The height of the pallet conveyor / FLT interface must align to the FLT requirements.	3. DD	Will this be provided?	
Op 6	Visibility of normal conditions	Threading the hooder not easy.	3. DD	How will this be improved?	
Op 8	Workstation design	All operator controls on the inside in central area to provide line of sight.	3. DD	How will this be provided?	

FIGURE 4.10
Checklist development.

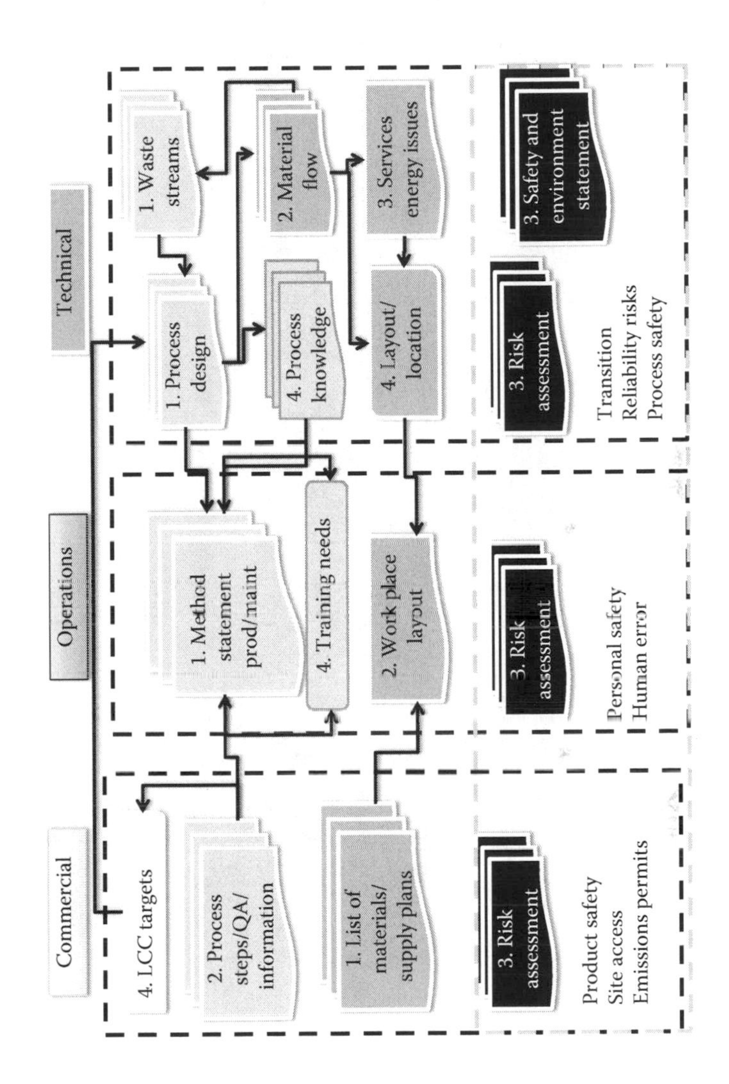

FIGURE 4.11
Module review outputs.

TABLE 4.1

Business Process User Specification

	Section	Contents
1	Aims	System vision, aims, and targets. What problem are we trying to solve? How do we want the future to be?
2	Principles	The rules by which the system will operate.
3	Ways of working	What will be done and when (daily, weekly, monthly, annually)?
4	Systems and procedures	The systems, files, and procedures that underpin the ways of working.
5	Responsibilities	Who will create/amend/operate the systems and procedures?
6	Next steps	What is the status of the specification and the next steps in its development?

Use these headings to develop the systems specification through the stages of

- Draft/concept specification
- Systems HLD
- System detailed design and forward program
- System configuration and user manual
- User training, systems validation, and go-live

4.2.5.1 Draft Specification

Concentrate on defining the "Aims" and "Principles" sections to develop a draft specification. Include outline ideas for Sections 3 to 5 (Figure 4.12).

Use this to brief team members and support an assessment of current systems to identify weaknesses or areas that require modification. Where the project involves the purchase of software, use the draft specification as a basis for short-listing potential software offerings.

The outputs from the draft specification step should include

- Aims and principles
- Agreed areas of weakness
- Gaps in current systems
- Potential next steps

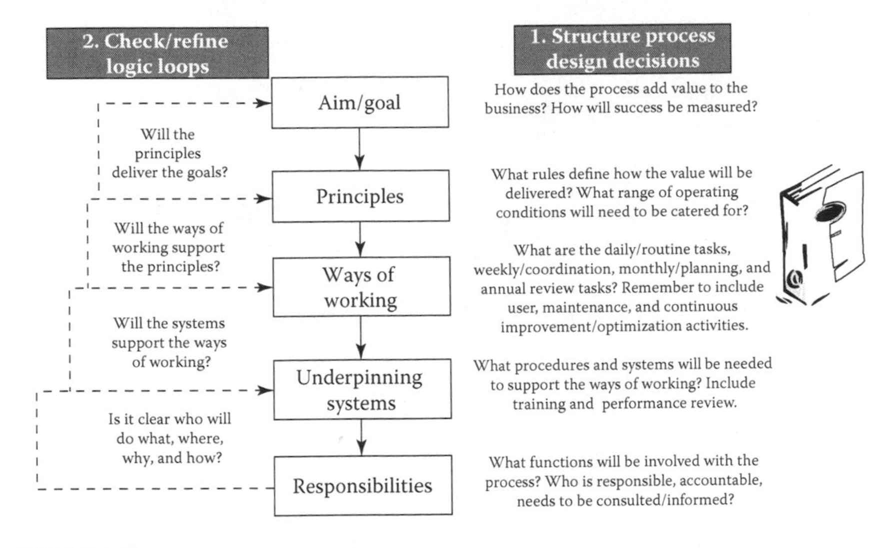

FIGURE 4.12
Resource management process specification.

4.2.5.2 Systems High-Level Design

After the circulation of the draft specification, add detail to the ways of working and systems and procedures sections. Where existing systems are failing, consider using pilot projects to identify why it is not working and how it can be improved. This will help with the specification and implementation of any new systems as well as provide a true benchmark for comparing alternative approaches.

As part of this step, generate a short list of practical options to deliver the aims and principles. Agree a basis for selecting a preferred approach from the short list. This may include testing options or simulating different ways of working to confirm benefits.

The HLD specification should include

- An agreed specification of the preceding six sections
- An outline implementation plan
- Risks to be managed (from commercial, operations, and technology [IT] functional perspectives)
- Proposed next steps in the form of a milestone plan

4.2.5.3 Systems Detailed Design and Forward Program

During this step, develop the contents of the specification document in the following sequence.

- Section 4, "Systems and Procedures," then from that identify how the system will be used in your organization to populate the ways of working section
- Section 3, "Ways of Working," can then be used to define roles and responsibilities and from that user manuals/training documentation
- Section 5, "Responsibilities," section can then be refined into an organizational structure so that job roles and management responsibilities can be assigned to individuals
- Section 1, Once the above sections are complete, revisit and refine "Aims," and Section 2, "Principles," into policy statements to be inserted into relevant policy documents/quality systems
- Section 6, after this pass through the specification headings any action points can be summarized under "Next Steps" and formatted into a forward implementation timetable

The outputs from this process are systems configuration, implementation plans, draft user manuals, and training plans.

4.2.5.4 User Training, Systems Validation, and Go-Live

Systems validation is a systems QA process that makes sure the system can support the ways of working set out in the specification. It is also used to stress test the system and ensure that it can handle real-life data and operating conditions. This is the systems equivalent of witnessed inspection. This is most effective when carried out by future system users following draft ways of working to test system functionality.

Preparation for go-live can often include the need for data cleaning prior to data migration. There are two types of data to consider: *static data*, such as contact details or bills of materials, and *dynamic data*, such as customer orders and stock levels. During this process, capture the lessons learned and update the specification, user manuals, and training plans.

Outputs from this step include the transfer of ownership for the new system so that the new owners control its evolution in line with the business's needs.

4.2.6 Setting Design Targets

As a measure of good design, LCCs are hard to beat. LCCs consist of capital costs and operational costs. To deliver a design that achieves the target of lowest possible LCC, the design must

- Provide what customers want at a profit (customer value and cost)
- Be easy to use and to look after (operability and maintainability)
- Produce output with low risk and high consistency (safety and reliability)

These characteristics are used as the basis for setting the six EEM design goals shown in parentheses. An explanation of each goal and a scale of assessment is included in Appendix A.

When these goals are used to assess option strengths and weaknesses they provide a relatively simple measure of LCC capability and therefore design effectiveness.

As in the case of the objective testing case study from Chapter 3, a more detailed cost analysis is often needed to compare short-listed options and explore opportunities to reduce LCCs and increase project value. This involves the use of simple LCC models to identify the main contributors to capital and operational costs.

The creation of the LCC model means that cost can be considered as a variable in the same way as any other project parameter. With it we can explore how to achieve the best value for the lowest cost. This approach of treating cost as a design variable was first developed for investments in military applications, public utilities, and not-for-profit business organizations to understand the trade-offs between the levels of capital costs and project value. The outcomes from this approach are graduated design targets linked to LCC cash flow estimates. It is then possible to assess the marginal returns from investing more or less capital. It can also be used to specify what is inside and outside of the scope of the project as a lever against scope creep.

In most cases, a simple LCC model will be sufficient to evaluate options and select a preferred approach. Simple models are used to provide a basis for comparing the impact of design options and features rather than to produce an accurate forecast of total LCCs. Typically, this involves the creation of base-case cost models using capital costs and operating cash flows for around five years. The cash flows for each option are then compared with the base case to identify the difference in cash flows from the base case. (An example of how this is used is included in Section 4.2.6.) Five years is within the time line for strategic planning and is therefore sufficient for this analysis. In most cases, extending the analysis time line will not increase the value of the result.

When creating the base-case model align the assumptions and parameters used with those applied by other business-planning processes, including

- Site business plans
- Capacity planning models
- Budgets and material costing

Simple LCC models are used to support EEM step decisions, as set out in Table 4.2.

TABLE 4.2

Uses of LCC Modeling

	EEM Step	Goal	LCC Application
1	Concept	To define the project scope	To compare scenarios and assess the likely benefits of the preferred option
2	HLD	To clarify the delivery approach, obtain funding, and select the right partner	To compare delivery approaches with the preferred option and firm up the EEM targets
3	Detailed design	To tease out latent design weaknesses, PP, enhance project value	To analyze the main contributors to LCCs and improve project, design, and operational efficiencies
4/5	Installation/ commissioning	Delivery of flawless operation	To refine operating methods and support problem-solving and focused improvement
6	SAT	Optimization	To confirm the achievement of LCC targets/benefits and support the development of focused improvement goals

The following is a typical RACI profile for the development and use of an LCC model. In this chart, *responsible* = set policy guidelines, brief, and support/ensure that the actions are completed to the required standard; *accountable* = complete specific actions; *consulted* = available to provide advice/direction as required; *informed* = keep advised on goals, plans, progress, the basis of decisions, and results/decisions made (Table 4.3).

TABLE 4.3

RACI Chart for LCC Model Application

	Task	Responsible	Accountable	Consulted	Informed
1	Base-case LCC model design	SGT	Fin	PM	ECT
2	Populate model	Fin	ECT	SM	PM
3	Approve/validate model	SGT	Fin	SM	ECT
4	Generate options and evaluate	PM	ECT	SM	Fin
5	Review and confirm analysis	Fin	PM	SM	ECT
6	Assess impact of strategic change on LCC model	SGT	PM	SM	ECT
7	Revisit/revise past project analysis as required	PM	ECT	Fin	SM

Key: Fin – finance, ECT = EEM core team, PM = project manager, SGT = stage gate team, SM = site management

4.2.6.1 Process Milestones

1. Create base case LCC model and develop options for review
2. Compare options and develop a preferred approach
3. Confirm/refine the preferred approach
4. Update LCC model and submit as part of stage gate review

A more detailed LCC analysis may be required, for example, to

- Support a fund application
- Review decision sensitivity to best-case/worst-case business scenarios
- Gain a greater insight into risks/opportunities

4.2.7 Basic LCC Model Example

In Chapter 3, the EEM core team compared the merits of option 2, an integrated filler labeler, and option 3, a separate filler/labeler solutions. The first step of the objective testing process identified option 2 as the preferred option even though this was considered to have a higher cost. The project LCC model was used to test that assumption as part of the second objective testing step.

The first part of the LCC analysis is to identify the parameters to be modeled and to collate the cost information for each option. This is set out in Table 4.4.

TABLE 4.4

Defining LCC Comparison Parameters

Parameter	LCC Analysis Approach	Action Needed	Parameter	Option 2	Option 3
Space needed	Building cost per m^2	Compare footprint size, adjust capital cost	Capital cost		+£440 k
Energy	Energy cost per case	Estimate percentage energy saving vs. benchmark	Energy	−15%	−15%
Waste	Frequency of jams × mat waste per jam + recovery time	Use same standard for both options based on line 3 benchmark	Material waste	.6%	.3%
Labor	Labor cost per case	Estimate percentage difference in team size from benchmark	Labor cost		+£50 k pa
Changeover	Increased line availability, impact on capacity	Team labor cost per run hour by time saving	Labor cost		+£1.8 k pa
Capital cost	Most recent cost	£1 m differential	Capital cost	£730 k	

Base case	Capex £k	Year						
		2016	2017	2018	2019	2020	Total	%
Capital costs	8200.00	0.00	0.00	0.00	0.00	0.00	8200.00	18.9%
Operating costs	0.00	7018.55	7018.55	7018.22	7018.55	7018.55	35092.42	81.1%
Total annual cashflow	8200.00	7018.55	7018.55	7018.22	7018.55	7018.55	43292.4212	
Cumulative cashflow	8200.00	15218.55	22237.10	29255.32	36273.87	43292.42		

Separate labeler LCC base case +£537k

Option 2	Capex £k	Year						
		2016	2017	2018	2019	2020	Total	%
Capital costs	8200.00						8200.00	18.7%
Operating costs		7125.94	7125.94	7125.61	7125.94	7125.94	35629.35	81.3%
Total annual cashflow	8200.00	7125.94	7125.94	7125.61	7125.94	7125.94	43829.3525	
Cumulative cashflow	8200.00	15325.94	22451.87	29577.48	36703.42	43829.35		

Integrated labeler LCC base case −£108k

Option 3	Capex £k	Year						
		2016	2017	2018	2019	2020	Total	%
Capital costs	7930.00						7930.00	18.4%
Operating costs		7050.96	7050.96	7050.63	7050.96	7050.96	35254.47	81.6%
Total annual cashflow	7930.00	7050.96	7050.96	7050.63	7050.96	7050.96	43184.4702	
Cumulative cashflow	7930.00	14980.96	22031.92	29082.55	36133.51	43184.47		

	Base case £k	Option 3	Option 2	Option 3-2
Capital costs	£ 8,200.00	£ 7,930.00	£ 8,200.00	−£ 270.00
Material	£ 27,008.15	£ 27,170.20	£ 27,089.18	£ 81.02
Energy	£ 2,830.63	£ 2,830.63	£ 2,830.63	£ –
Production	£ 4,045.34	£ 4,045.34	£ 4,501.25	−£ 455.91
Maintenance materials	£ 1,208.30	£ 1,208.30	£ 1,208.30	£ –
	£ 43,292.42	£ 43,184.47	£ 43,829.35	
			Option 3	−£ 644.88

FIGURE 4.13
LCC model example.

The base-case LCC model was adjusted using this data to compare the LCC cash flows for each option with the base-case model (Figure 4.13).

The data for option 2 produced an LCC £108,000 lower than in the base-case model. Option 3 produced an LCC of £537,000. This analysis reversed the initial working assumption that, based on guide prices, option 2 was more expensive. This strengthened the assessment of option 2 as the preferred option. In addition, as the base-case model included the operational savings on which the project was justified, the option also increased the expected return on investment.

As the base-case LCC model had already been created, this comparison took less than 30 minutes to complete as part of a core team session.

4.2.8 Estimating Costs

4.2.8.1 Operating Cost Estimates

The base-case LCC model in the preceding example was built using capital cost estimates from an approved capital-funding application plus operational budget data adjusted for future years based on forecast output levels. These costs were modified to account for anticipated changes in material usage, performance targets, and projected energy savings. The model took around half a day to build and validate.

4.2.8.2 Capital Cost Estimates

There are many capital cost estimation methodologies and a full exploration of the merits of each is outside of the scope of this book, but the following are a number of options to consider:

- *Vendor cost estimates* are the simplest basis, but reliable estimates may not be available prior to the formal ITT process.
- *Plant cost indices* such as the Chemical Engineering Plant Cost Index publish monthly and annual indices of cost trends for process plant construction. This is made up of 11 individual indices covering a range of categories from pumps, valves, and fittings to process instrumentation. They suggest that their indices can be used with confidence to predict costs over a five-year period. This means that you could look back at past costs and use their index to estimate current investment for similar assets. The index is published in the monthly.
- *Chemical Engineer* magazine a couple of months after their analysis. More timely data is available for subscribers. Other indices are also available.
- *Lang factor* models apply factors to the cost of main plant items to estimate the total cost. The model shown in the following example uses seven Lang factor parameters.

1. Installation
2. Piping
3. Instrumentation
4. Electrical
5. Civil
6. Structures and buildings
7. Lagging

The Lang factor parameters are adjusted to take into account the nature of the project. In the example shown, the factor selected for installation is 1.1, a multiple used for projects where much of the site erection is included in the purchase price. Here the installation cost is estimated at 1.55 times the main plant item cost. The model also includes an estimated design cost of 25% of the main plant item costs (Figure 4.14).

These estimates are made on a boundary limit basis. That is, they do not include any additional costs outside of the boundary of the project—for example, costs for site clearance or the provision of additional site utility capacity.

Basis of capital estimate: Working sheet

Equipment	Compressor	Version	Date
Notes	Example of using sheet	1.0	dd-mm-yy

Lang factor estimate

Compressor project	£k	Cost category
Main plant items	100	4
Estimated installation costs	155	
Direct plant costs	255	
Design/overheads	0.25	
Total installed cost	£319	k
Site costs		k
Contingency		k
Grand total	£319	

Lang factor estimate detail

	Description	Applicable	Factor	Cost £k	Notes
1	**Installation**		0.19	19	Selected midpoint between 1C and 1D
1	Much of site erection included in purchase price	y	0.03	3	Minimum site fabrication but alignment
2	**Piping including installation**				Selected lookup table standard
2	Large bore piping with complex system	y	0.41	41	
3	**Instrumentation**				Selected lookup table standard
3	Controllers and instruments	y	0.33	33	
4	**Electrical**				
4	Lighting and power excluding transformers and switchgear for machine main drives (e.g., pumps, compressors, crushers)	y	0.18	18	
5	**Civil**		0.12	12	Selected midpoint between 5A and 5B
6	Plant in a simple covered building	y	0.29	29	
7	**Lagging**				
	Estimated installation cost		1.55	155	

FIGURE 4.14
Lang factor model example.

4.2.9 ITT Design

The content of an ITT or request for quotation (RFQ) pack varies from company to company. Figure 4.15 shows an example of an ITT structure.

Background information

- Overview
  - Project scope, deliverables, limits and exclusions
  - Safety standards,
  - Project context, rational, justification and impact
- Products and capacities
  - List of products with pack and unit load details including v graph data)
  - Buffers (accumulation and material stack times)
- Utilities
  - Energy consumption. Guarantees consumption/usage rates per unit of output
- Training
  - Timetable
  - Training cascade and timing plan
- Layouts

Equipment requirements

- Equipment standards
  - Design module definitions
    - Capacities, working methods, asset care
  - Design guidelines
    - Safety, Reliability, Operability, Maintainability, Customer value, Life cycle cost
  - Technical standards
    - Change parts, visual management
    - Spare parts
    - Lubrication, Automation, Electrical, instrumentation/calibration requirements, Piping materials, Cleaning systems.
- Cost summary headings
  - Equipment, Training, General services (resources for Engineering, documentation, transport (FOB), Packaging, Installation, Commissioning, SAT)
  - Legal compliance
  - Project delivery costs (spares, travel etc)

FIGURE 4.15
Tender specification headings example.

Include a vendor EEM self-assessment questionnaire covering EEM design goals and guidelines to

- Assess vendor appreciation of the EEM benchmarks of safety, reliability, operability, maintainability, customer value, and LCCs
- Raise vendor awareness of design goals, guidelines, and standards—both an aid to vendor selection and setting the foundation for a future collaborative working relationship

On return of the completed questionnaire, the vendor scores are calibrated by the core team based on the evidence provided. This is then used as an assessment of design fit to support the comparison of vendor offerings.

4.2.9.1 Developing Design Guidelines

- If this is a first project, create around 10 design guidelines based on the module review outputs. This list should set out broad design features (Figure 4.16).

	Vendor EEM questions list
Site	Princes limited eden valley
Asset	Line 4
Vendor	

Instructions

- A list of EEM standards have been developed and will be considered throughout the entire length of this project.
- Please answer the questions in the Question column.
- If YES, please go on to tell us how well you believe your design meets the question in Column I (1: barely meets, 2: partly meets, 3: fully meets, 4: has a relevant reference site, 5: can take you to 3 or more).
- If NO or N/A, please go on to tell us why in Column I.
- If you wish to supply any supporting information against your answer, please do so in Column J.

Code (A)	Standard (B)	Scope (C)	Module (D)	Weakness/ issue (E)	3. Acceptable standard (F)	Question (G)	Answer (Yes, No, N/A) (H)	Degree to which you comply (1–5) (I)
Rel. 1	Changeover SMED	Mold change over time, low hopper alarms, auto-splice, auto check, accumulation lane changes	General	Minimize the need for intervention	Minimum manual activity, no tools required	Does your design support minimal changeover activity without the need for tools?		
Rel. 2	Material usage	Material quantities and reconciliation, blower auto change, confirmation of materials	General	Reduce risk of human error and admin needed to maintain traceability of materials	Counters at key points to capture material usage/rejects	Does your design include counters for good and scrap materials at each production step? Is it easy to capture and archive this information?		
Rel. 3	CIP/sanitization	Sanitization and utilities	General	Minimize build up of contaminants, reduce sanitation overhead	Optimized CIP time and frequency to achieve zero microfailures	Does your design include optimized CIP as per standard?		
Rel. 4	Workplace organization	Energy efficiency, storage, wash facilities, networking, mods to existing lines, tracking label production, CIP split, guide rail adjustment, operator levels	General	Minimize movement, allow at least 30 minutes between intervention	Formal layout with clearly defined work space change parts, communications, quality checks, and materials integrated with other lines	Does your layout include recommended space requirements for the items set out in the acceptable standard?		
Rel. 5	QC checks	Auto-checks and reject handling, how to reduce QC checks by 1/2	General	Minimize need for manual checks to assure quality	Right manual process controls in the right place	Does your process include comprehensive product quality defects check and reject features?		
Rel. 6	Protection	Protection of components from damage, and accelerated wear	General	Items at risk of damage from fallen items, water/chemicals, or contamination	Sensitive/easily damaged items are protected from accelerated deterioration	Are components protected from accelerated wear, impact damage, or liquid damage?		
Rel. 7	Jams	RM, product, and palletized loads	General	Minor stops causing loss of performance when running	Conditions set to minimize jams, early problem detection to identify potential jams before they result in a production outage			

FIGURE 4.16
Vendor EEM questionnaire.

- As can been seen from the questionnaire example, vendors have space to add examples/evidence of compliance or provide information where they are not able to meet the standard.
- Following funding approval, circulate the questions to the vendor short list.
- Evaluate vendor response against the six targets and identify topics to discuss with vendors/probe their ability to comply (Figure 4.17).

Vendor 1			
Intrinsic safety	3.86	3.21	2.79
Intrinsic reliability	2.57		
Operability	2.57	2.50	
Maintainability	2.43		
Customer value	2.71	2.64	
LCC	2.57		

Vendor response calibrated by evidence supplied prior to further discussion with short listed vendors

Vendor 2			
Intrinsic safety	2.57	2.64	2.43
Intrinsic reliability	2.71		
Operability	2.00	2.14	
Maintainability	2.29		
Customer value	2.57	2.50	
LCC	2.43		

Vendor 3			
Intrinsic safety	2.43	2.43	2.26
Intrinsic reliability	2.43		
Operability	2.57	2.71	
Maintainability	2.86		
Customer value	2.14	1.64	
LCC	1.14		

Vendor 4			
Intrinsic safety	2.57	2.43	2.19
Intrinsic reliability	2.29		
Operability	2.57	2.29	
Maintainability	2.00		
Customer value	1.86	1.86	
LCC	1.86		

FIGURE 4.17
EEM vendor comparison.

- Use these responses to aid vendor short-listing prior to calling in vendors for a follow-up meeting.
- Finalize the evaluation and select a preferred vendor and agree contractual terms, then place the order.

4.2.9.2 Process Milestones

1. A set of questions to assess vendor ability to provide EEM benchmark performance
2. Responses from vendors to EEM benchmark questions
3. Evaluation of vendor responses and follow-up as necessary
4. Selection of preferred vendor, contractual terms, and order placement

4.3 DETAILED DESIGN

The detailed design process adds detail to the specification documents developed during the HLD step. This includes layouts, module review outputs, and resource management specifications (Figure 4.18).

Often vendors will want to get detailed drawings completed quickly so that they can start cutting metal as soon as possible. The drawing office can often be a bottleneck and on the critical path to achieve their order fulfillment process time line.

This pressure should be resisted until it is clear that the vendor team understand what is required. This avoids the risk that time is wasted on design revisions later. The vendor will have allocated a budget to this process, and their internal cost control will mean that once the allocation of vendor drawing office resources have been used up, changes will be resisted.

EEM detailed design uses a two-stage process, as set out in Figure 4.19.

- Meeting 1 involves the assessment of an outline design using 10%-outline design drawings to explain the EEM goals and guidelines, make clear the operational realities, and outline ways of working. This will bring detailed design priorities to the surface. The vendor then uses this information to produce 90% drawings, which confirm how those design goals will be met.
- Meeting 2 involves the vendor explaining in detail how the more detailed design meets the design goals and the forward program to deliver flawless operation.

	Document	Description/Notes
	Layouts	
1	Location/space allocation	Top-level drawing showing main aisles, work areas, and access/safety hot spots. Work area layouts showing draft workplace layout and content
2	Material and workflow	Input and output volumes and routes for each module covering RM, Waste, Finished goods. Identify conveyor, mechanical handling, and manual handling routes and volumes
	Process flow charts	
3	Process steps and outputs	From module review page 1
4	Process controls/working method	Module review page 1
5	Material supply and transfer to line	Coordination for point 2 above
6	Product production plan (plan for every product)	Working assumptions re: production planning capacities, SKU range, batch sizes, etc.
	Supporting docs	
7	Energy management	Identification of potential areas for energy reduction
8	Process knowledge, skills, and working methods	Identification of gaps and areas where we need to raise current standards
9	Transition management	Tactics to assure customer service is maintained on existing lines and how will the new line capacity will be ramped up.
10	FAT Protocol issues/targets	Where are the construction and installation risks and quality assurance review points?
11	SAT Protocol issues/targets	What are the performance targets to be achieved?
	Ways of working	
12	Start-up steady-state close-down	From module review
13	Asset Care Plan (SCRIPT)	Where are the opportunities for maintenance prevention?
14	Working environment issues	Identification of potential hazards and preferences
15	Organization and skill profile	Working assumptions re: task, role and organizational
	Resource management	
16	Material movement and control	How materials will be controlled
17	Services	What services are required
18	QA Processes/defect reduction	What QA processes are required? Where can we reduce the root causes of quality defects?
19	Life cycle cost model	What are the main contributors to LCC?

FIGURE 4.18
Preparation for detailed design.

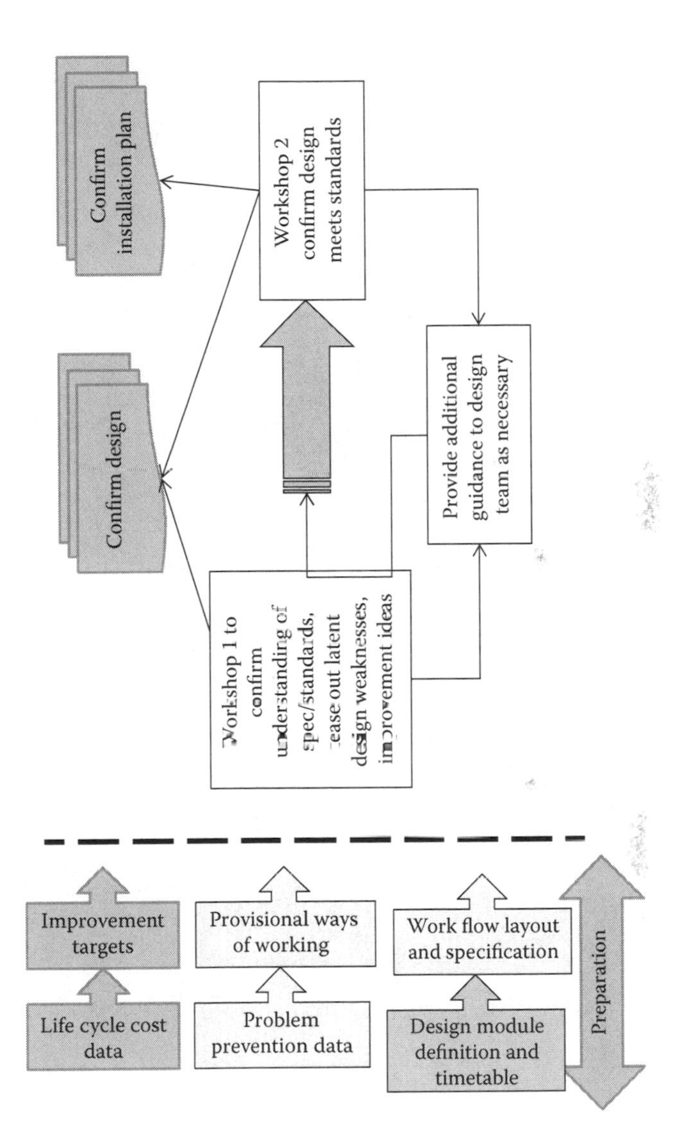

FIGURE 4.19
Design workshop preparation.

4.3.1 Module Review Update

As part of the detailed design process, update the module review outputs, including the following:

- Complete a module criticality assessment.
  - The identification of *maintainability, operability, reliability* and *safety*, and *environmental* risks
- Complete a health check to confirm current standards of operational practices, asset plans, and maintenance effectiveness.
- Complete the stability and resilience assessment as a team under the following headings:

  - Confirm Operational stability
    - Operations best practice (start-up/changeover, steady state, and close-down/clean-out)
    - Preventive maintenance (cleaning, inspection, lubrication, and tightening)
    - Servicing (restore deterioration)
  - Confirm Technical resilience
    - Measuring deterioration (condition monitoring)
    - Spares management
    - Failure protection (spare capacity, warn, relieve, shut down, prevent)

Review the residual risk scores to identify priorities for the development of new practices and current practices in need of improvement. The specification documents will then provide information in a format to

1. Explain how the vendor technical specification will deliver EEM goals and target performance
2. Support project QA through a witnessed inspection program
3. Manage the glide path to flawless operation including internal operational change, installation, and commissioning

4.3.2 Dealing with Risk

4.3.2.1 Specify Low-Risk Solutions

The purpose of a risk assessment process is to identify potential hazards and apply hazard controls to reduce risk to an acceptable level. Table 4.5

sets out a hazard control hierarchy, referred to by the acronym ERICPD, designed to support the mitigation of risks from operational hazards. Chapter 5 explores how the logic of this hierarchy of control levels is equally relevant to minimize other areas of risk such as financial or market risks.

TABLE 4.5

Hazard Control Hierarchy

Hazard Control Level	Example
Eliminate causes of risk	Restore or assure basic equipment conditions, standardize/simplify working methods to remove human error risk, use visual indicators to assure safe setup/operation of equipment, training; noise reduction, removal of sharp edges; safe design
Reduce impact of risk	Equipment guarding, capture of hazardous materials/bunding, blast protection, dousing showers, escape routes
Isolate causes of risk	Segregation of chemicals or contaminants
Control working methods	Safe systems of work/work permit controls for nonroutine tasks
Protect from risk	Personal protective clothing and equipment
Discipline	Safety audits, behavioral safety

EEM design guidelines/standards include definitions for acceptable standards to deal with common hazards, including those relating to access, ergonomics, utilities, equipment, height, isolation, line of sight, material flow, microbiology, product safety/QA, and statutory compliance. An assessment of HLD using these EEM guidelines/standards is sufficient to support concept and HLD risk assessment. This includes the specification of expectations to vendors as part of the ITT documentation.

A more detailed risk assessment is carried out prior to completion of the detailed design stage gate review to

- Confirm the elimination of risk areas within the design where possible
- Explore cause–effect mechanisms of potential risks and agree problem prevention tactics
- Outline working methods and installation, and commissioning/operational handover plans
- Define safety protocols
- Confirm installation/precommissioning and plan commissioning activity

The outputs from this are captured as part of the design module specification, as shown in Figure 4.20. Risk assessment is covered in more detail in Chapter 5.

Process	Best-practice operations	Maintenance
Inputs (including quantities/typical batch size) • Materials • Services	Start-up including set up	Routine checks/prevention
Process steps • Functions/zones • Critical parameters	Steady state (including typical problems/ interventions)	Inspection
Outputs • Product • Waste	Close-down and clean out	Servicing
Gaps/issues/concerns/ opportunities	Gaps/issues/concerns/ opportunities	Gaps/issues/concerns/ opportunities

FIGURE 4.20
Design module specification.

4.3.2.2 Specify Problems to Be Prevented

Use the specification development process to support the adoption of PP principles during the design process by providing

- An explanation of the shop floor reality/PP data
- Potential sources of defects
  - Tooling, critical-to-quality components, control of optimum conditions, reducing causes of accelerated wear
- Good-practice examples of operability and maintainability
  - Methods
    - Easy-to-establish reference planes, easy to clamp, easy to position in fixtures, go/no-go visual adjustment protocol
    - Normal conditions visible at a glance, red/green limit indicators—for example, liquid/gas flow/pressure, mechanical pressure rollers/scraper, pump flow/pressure, electrical loading, minimum/maximum levels, lubrication types
  - Materials
    - Resist damage, rarely form burrs, easy to machine, easy to measure

- Machinery
  - Contains, reduces, eliminates the scattering of swarf, dust, or dirt; easy to start up, change over, run, and close-down; minimal cleaning with no hard-to-clean places
  - Access/ease of inspection, early problem detection, routine servicing
- Manpower
  - Easy to distinguish from other products or components, easy to assemble, easy to automate
- Visual workplace, layout/signage, labeling, and color-coding standards
- Visual management policies for routine information, work instructions, and workflow synchronization (Figure 4.21)

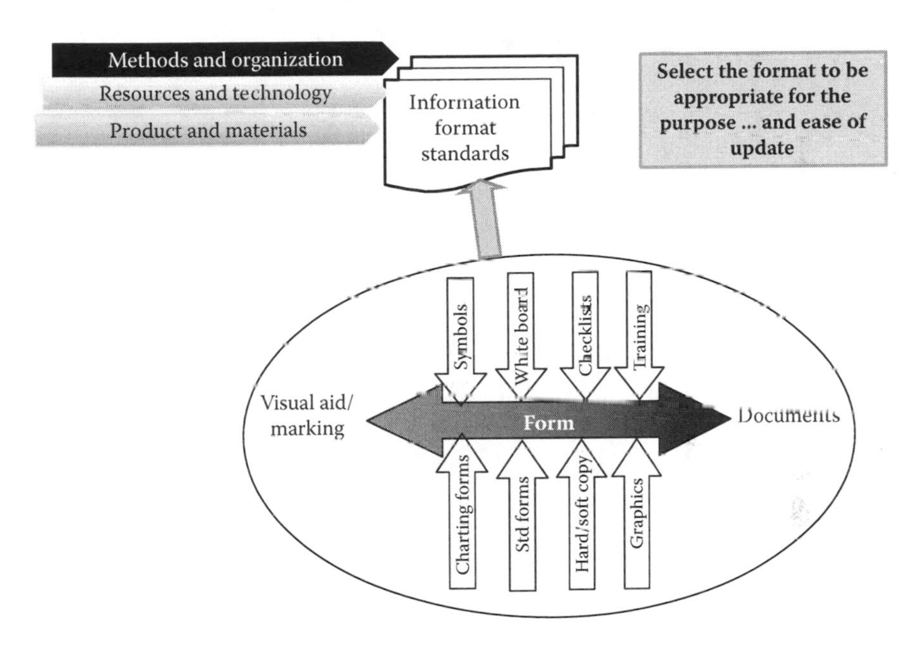

FIGURE 4.21
Documentation and visual format standards.

4.3.2.3 Use Charts and Graphs to Raise Awareness

Where testing is carried out to confirm process cause–effect mechanisms, record insight gains in a format that aids communication and share lessons learned using charts and diagrams such as X-charts (Figure 4.22).

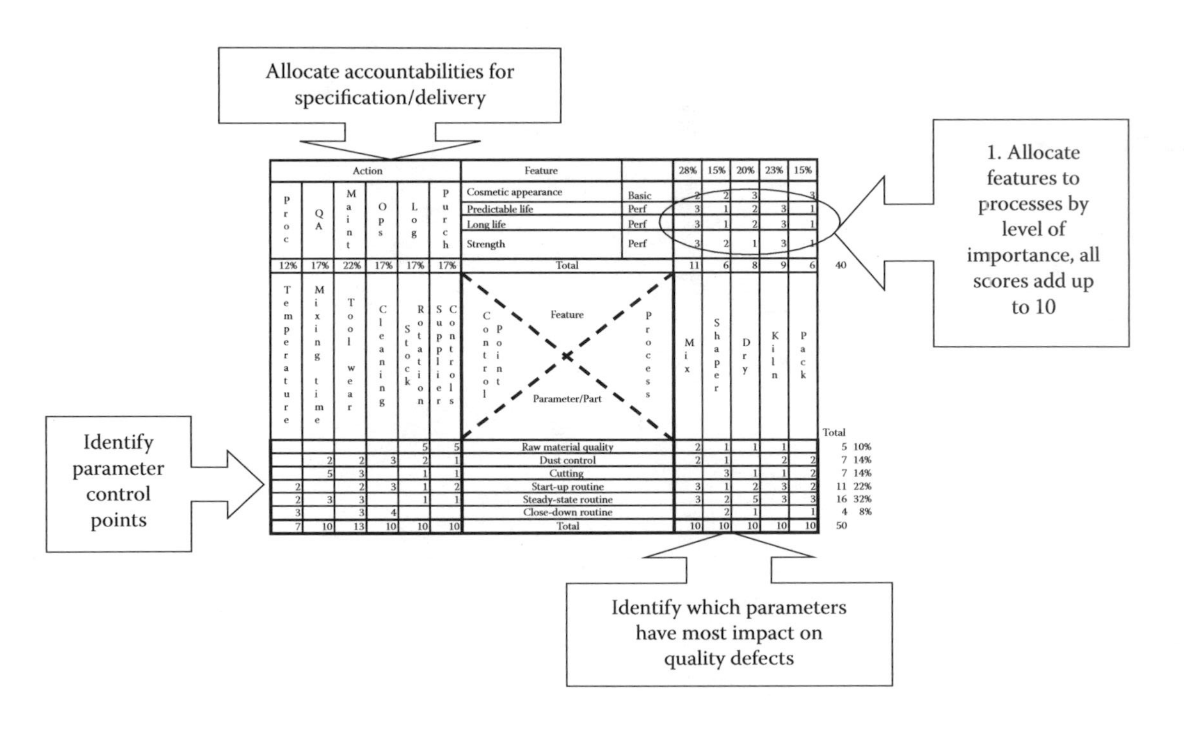

FIGURE 4.22
Record-testing insights gained.

4.3.2.4 Specify Learning Plans That Reduce Error Risks

Weaknesses in the skill development process are a common cause of error risk. To avoid this pitfall, specify skill development needs in a way that guides the learning process, covering

- Core/routine tasks (routine tasks requiring limited judgment—e.g., machine minding during steady state)
- Intermediate tasks (routine tasks requiring technical judgment—e.g., nonstandard adjustments, routine servicing)
- Specialist tasks (nonroutine tasks requiring technical judgment—e.g., troubleshooting, nonroutine servicing)

Set graduated skill development outcomes and goals—for example:

- Internal personnel are to be capable of completing core tasks unaided at the start of production day one.
- Intermediate skills are to be developed during stabilization.
- Specialist skill development will be part of the definition and delivery of optimum conditions prior to the site acceptance test (SAT).

These goals are incorporated in the training cascade shown in Figure 4.23.

4.3.3 Detailed Activity Planning

One of the outputs from the detailed design process is the forward plan, including

- The road map to develop core, intermediate, and specialist competencies
- Operating methods
  - Workflow, defect prevention, best-practice development route (*plan, do, check, act*, or PDCA), job/skill profiles, training/skill development
- Operations organization
  - Workplace information, work planning/scheduling, recording and reporting, cost control, stock management, operations organization

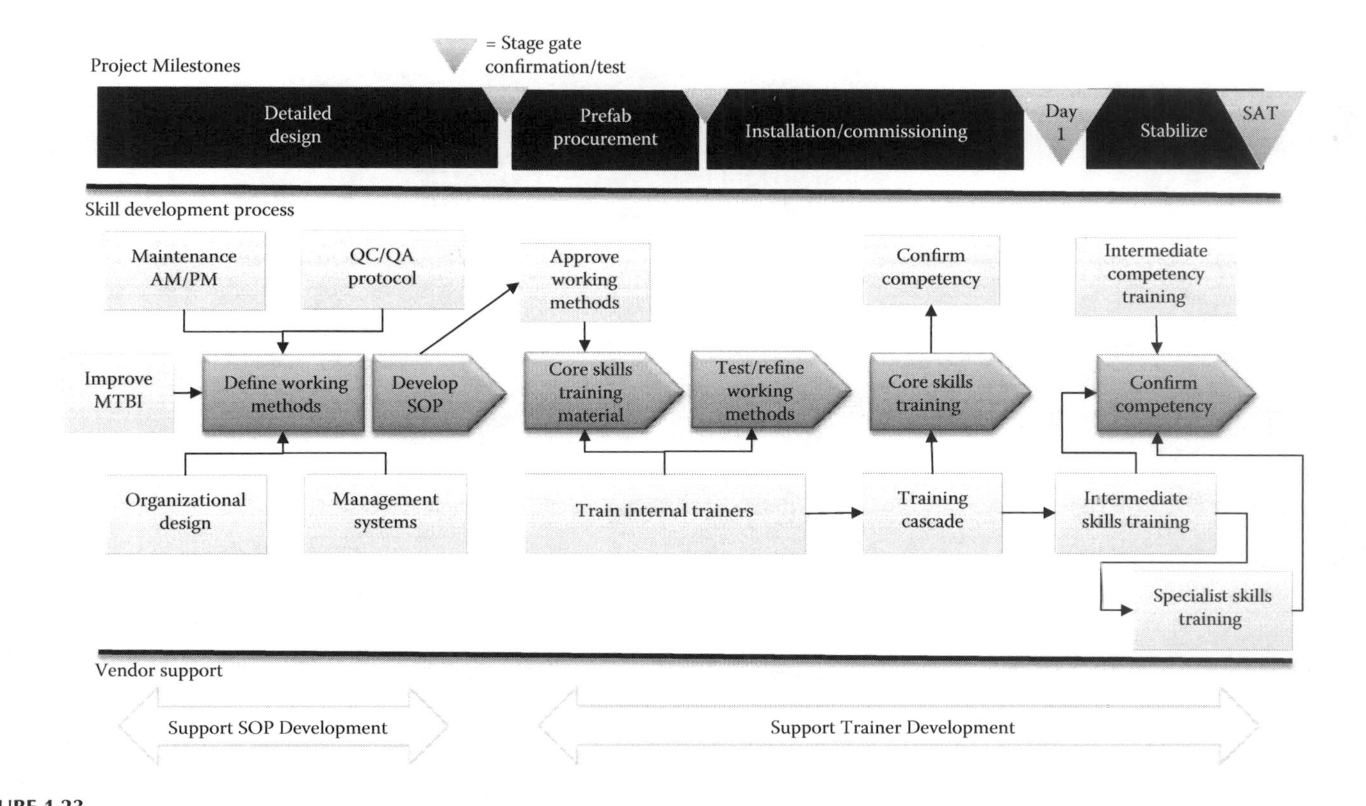

FIGURE 4.23
Skill development process.

- Maintenance methods
    - Maintenance prevention, best-practice evolution (PDCA), inspection, spare parts, machine manuals, service-level agreements, maintenance organization
- Maintenance organization
    - Technical information, planned maintenance, planning/scheduling, recording and reporting, cost control, lubrication, and spares control

4.3.4 Change Control

Once the risk assessments have confirmed the design and installation plans, the design is frozen and change management protocols are applied. At this point the project cost budget is updated to reflect changes in cost and targeted benefits.

Further design changes require authorization using a formal change control procedure involving assessment by the stage gate team, including

- Reasons for the change, references, and consequences for commercial, operations, and technical
- Cost and planning consequences

4.3.5 Witnessed Inspection

Anyone who has built flat-pack furniture will know that a key task at the start of the job is to confirm that all of the components for the build have been supplied. At a project level this is known as a *witnessed inspection*. The forward project plan includes a series of witnessed inspections developed to confirm progress against project quality milestones. This provides a safety net of checks to trap problems, deviations from the agreed design, and potential delays (Figure 4.24).

The frequency and timing of witnessed inspections depends on the nature of the project but could include tests covering

- Vendor manufacturing activities
    - Based on manufacturer quality plans
- Construction progress
    - Including confirmation of key dimensions and datum points

Module/Equipment		Palletizer			Project step				
Test name					Date completed			By	
Test no	**Test name**	**Step**	**Planned Date**	**Method purpose**	**Test materials and standards**	**Test criteria**	**Tester Training/Skills**	**Instrument/Test standard**	**Notes/Status**
PalDD1	Machine efficiency and speed	Detailed design	Jan-15	Agree commissioning standard and materials required.	Completed guarantee sheet as per agreed contract.	Formal definition of test criteria at PDI, installation, and commissioning steps.	Understand machine ratings and measures.	None	Formalize process to confirm guarantees at commissioning.
PalDD2	Changeover	Detailed design	Jan-15	Carry out table-top DILO simulation on changeover from 500 to 750 ml to support development of draft operating procedures.	Best-practice list and module review DILO. Agreed layout.		How to develop draft SOP.	Confirm methods meet EEM operability standards.	Aim to define core and intermediate skills and working methods process charts.
PPD6	Stretch-wrap layer tests	Detailed design	Jan-15	Define test process to confirm standard layers and waterproof wrap levels.	Customer specification.	Quantify materials for test run, agree location, date, and performance criteria.	Awareness of customer standards and moisture testing.	None	Validate method with customer representatative prior to testing.

FIGURE 4.24
Test summary extract.

- Factory acceptance tests (FATs)/predelivery inspection
    - Based on internal commissioning plans and validation testing
- Skill development
- Installation
    - Based on vendor installation quality plans
- Commissioning
    - Based on internal commissioning plans and validation testing
- SATs
    - Based on the status of the commissioning tests and action map measurable outcomes

The use of visual management to support the scheduling and completion of witnessed inspection tests makes it easy to monitor progress on the glide path to flawless operation (Figure 4.25). This also helps to coordinate test resources with installation and commissioning work packet progress.

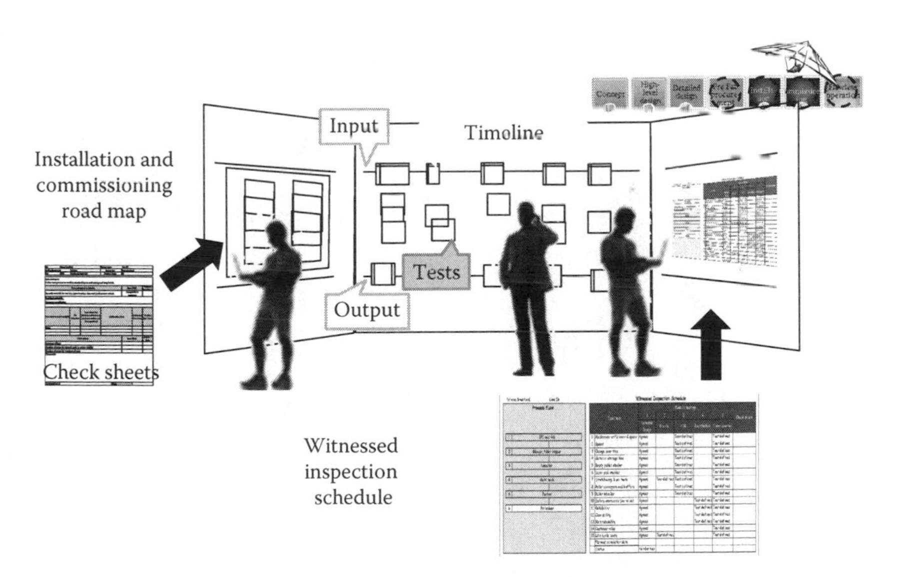

FIGURE 4.25
Witnessed inspection visual glide path.

4.4 DEFINE

Specification management during the concept and HLD steps involves activities to capture the following:

- The concept specification and benefits
- The HLD risk assessment
- The defined route to flawless operation
- The project justification and LCC model

Figure 4.26 provides a list of documents and processes commonly introduced during the define phase. These are also used later in the project.

4.5 DESIGN

Specification and LCC management tasks during detailed design and pre-fab procurement steps cover activities to

- Formalize the detailed project specification, LCC forecast, and forward plan
- Risk assess detailed design, installation, and commissioning plans
- Introduce engineering change management protocols
- Plan, organize, and control detailed installation, and commission plans with vendor/contractors
- Define and complete witnessed inspections
- Manage the construction phase, including witnessed inspection and PP activities
- Collate detailed design outputs into an operations specification covering
    - Issues/targets (including potential gains from the current benchmark asset)
    - Layout/material flow showing hot spots
    - An overview of the operation approach
        - Normal operation
        - Support for start-up, steady state, close-down/cleaning

	Purpose	Created first at	Used later for
1. Module definition	To provide structure for design development, collate information, and issues/concerns from module review.	High-level design	Procurement, detailed design, and development of operational working methods
2. Objective testing	To support comparison of option strengths and weaknesses	Concept	Decision support tool for all stage gates
3. Target setting	To clarify areas of defined and potential added value	High-level design	Once the final option has been selected, targets should be updated to take account of additional gains/added value to be achieved from the project
5. Life cycle cost model	To clarify cost drivers and support/refine option evaluation	High-level design	Feature selection, operational target setting, and confirmation of project delivery results
6. Vendor selection questionnaire	To support the evaluation of vendor offerings against EEM design goals and operational standards	High-level design	Used as part of evaluation of vendors and during detailed design discussions to assess design effectiveness and tease out latent design weaknesses
7. Witnessed inspection schedule	To summarize the project quality assurance tests/checks to confirm progress toward project goals	High-level design	Used/updated at all steps to site acceptance testing

FIGURE 4.26
EEM specification template summary.

- Daily maintenance
  - Lubrication, tightening nuts and bolts, deterioration check, minor servicing (e.g., replacing tools, suction cups, filters)
- Periodic maintenance
  - Inspection, testing, periodic servicing
  - Preventive maintenance
  - Trend testing, nonroutine servicing
  - Corrective maintenance
- Problem-solving, repair, process recovery
  - Equipment improvements
  - Strengthen, reduce loading, increase precision
  - Improved procedures, improved access, extended component life

4.6 REFINE AND IMPROVE

The specification and LCC management tasks during the installation, commissioning, and improvement steps cover

- Witnessed inspection, including day-one glide path readiness review
- The transition from installation to commissioning personnel
- Flawless operation delivery
- The specification of optimum conditions and improvement next steps
- Updating systems, procedures, and best-practice design books
- The transition from project to operation-focused improvement program
- Updating technical documentation
- Collating project Key Performance Indicator (KPI) analysis and confirmation (Figure 4.27)

4.6.1 Day-One Production and Site Acceptance Testing

When the witnessed inspection program for all modules has successfully passed the commissioning testing, approval is given to progress to initial operation status. This is the authority to operate the plant and begin steps to achieve full performance.

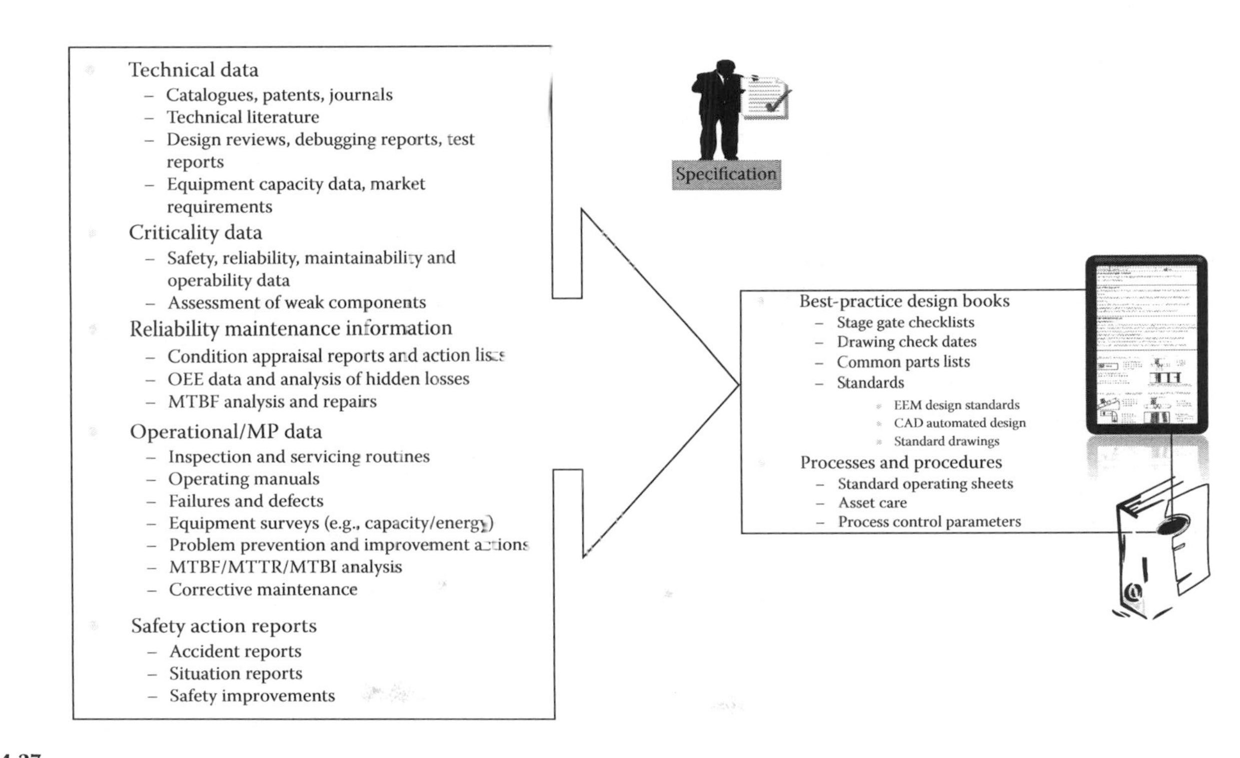

FIGURE 4.27
Operational documentation.

The SAT is completed when there is confirmation that the performance has passed the witnessed inspection stability tests. This typically covers

- Line performance
- Equipment performance
- Technological guarantees
- Utility consumptions
- Handover of up-to-date documentation, drawings, and manuals

4.7 SUMMARY

This chapter covers

- Making choices, refining designs, defining equipment functions, and design modules
- Documenting the preferred option to enhance project value
- Working with vendors

Key learning points are as follows:

- Build a collaborative relationship based on shared goals.
- Add detail to the specification sections at each step so that it evolves into an operations manual.
- Develop checklists based on issues raised to support the quality plan/witnessed inspection.
- Focus on reducing LCCs to get the best possible value from the investment in both money and brain power.
- Understand equipment losses and countermeasures and their impact on LCCs.
- Treat each project as a voyage of discovery using the LCC model as the hidden-value treasure map.

5

Project and Risk Management

This, the third of the five EEM subsystems introduced in Chapter 2, "The EEM Road Map," covers the planning, organization, and control of project resources to reveal and resolve latent problems/risks early and deliver project goals using the expertise of the complete internal and external team.

In addition to the strong links with design and specification, project managers typically are the lynchpin of the project governance process, providing guidance on

- Plan development
- Resource definition
- Timings and key decision points
- Project organization, accountabilities, and performance management
- Alignment of the relevant plans and activities
- The identification of next steps and actions to deal with evolving risks and roadblocks to progress

Project management involves the planning, organization, and control of activities to deliver business benefits (Figure 5.1).

Fundamental to the success of a project is the ease of interaction between project resources and between those resources and the day-to-day business. These aspects are part of the organizational elements of project management, which is why the task of project organization is rightly referred to as the *art of engagement*.

The final constituent of project management concerns control and coordination. In addition to feedback on progress and actions to stay on track, it also includes

- The repetition of key messages to assure retention
- The clarification of priorities and standards
- The management of bad news

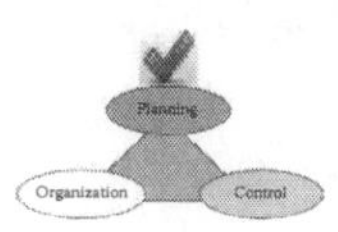

Planning: The art of resource management

- Confirm the specification/outcome, what do we want to achieve by when and to what standard
- Define project strategies, tasks, critical path/priorities, timetable, skills and resources needed.

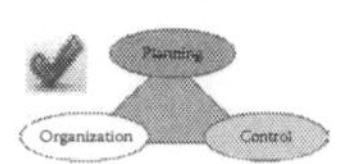

Organization: The art of engagement

- Define team roles and accountabilities
- Agree procedures, standards, and communications protocols with team and suppliers

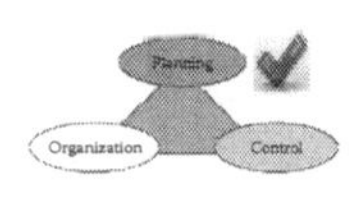

Control: The art of communication

- Mobilizing the project, confirm project brief, communicate with team and stakeholders
- Implement program resource management measure results, confirm progress, and learn from experience

FIGURE 5.1
Elements of project management.

In the latter stages of the project, this concerns managing the glide path to flawless operation in the same way as an airport control tower manages a plane coming in to land or a space agency landing a rocket on the moon. That is why project control can rightly be referred to as the *art of communication*.

This chapter covers best-practice project management, project team, and risk management processes before covering the specifics of the define, design, and refine project governance phases introduced in Chapter 2.

5.1 DEVELOPING PROJECT PLANS

A good project plan is both *realistic* and *achievable*—that is, realistic in terms of what the project needs to deliver and achievable in terms of the resources available. As resources are frequently the bottleneck to progress, the act of project planning can be reasonably defined as the *art of resource management*.

In the case of long-term plans, the resource being managed might be cash. In the case of short-term plans, the resource could be hours of available time. In this case resource management includes having the know-how to first decide what to do and then deliver it on time, on budget, and flawless from day one, as set out in Figure 5.2.

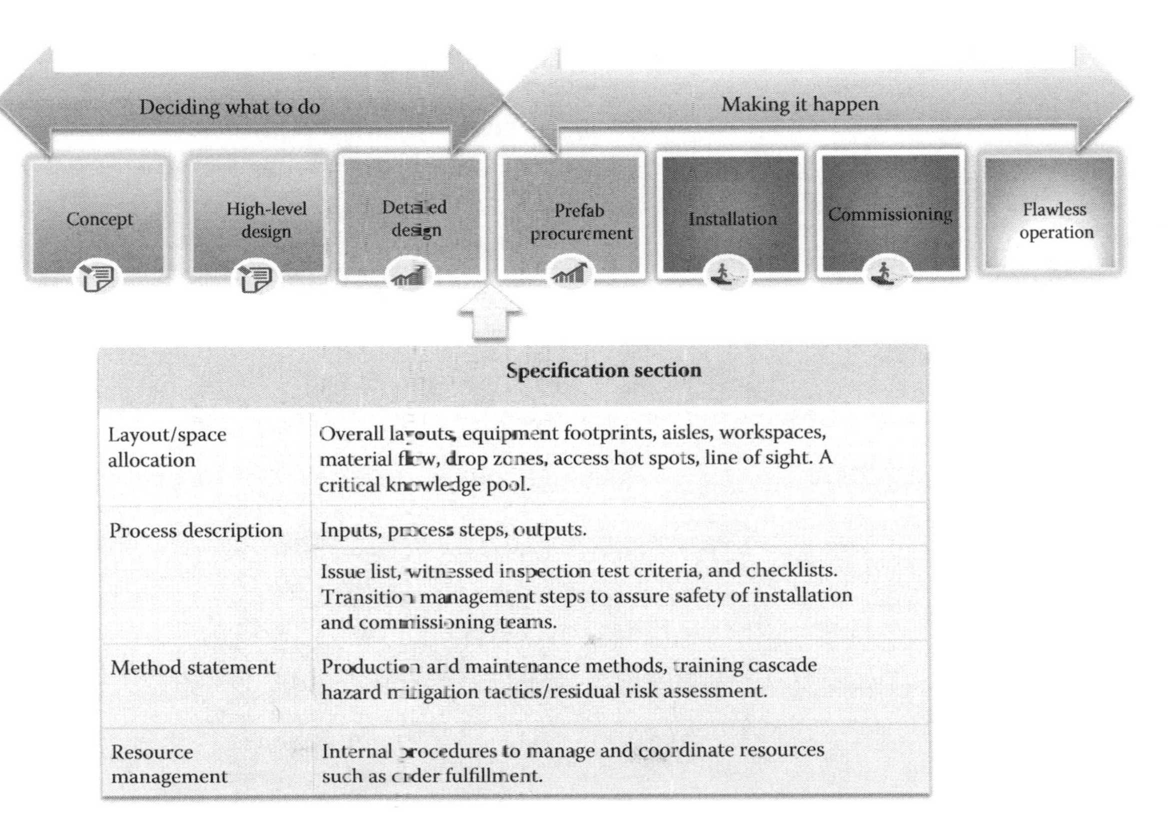

FIGURE 5.2
Project-managing delivery.

A common pitfall is to assume that a realistic and achievable project plan can be defined in detail at the start of a project. In reality this is not the case until after the detailed design is frozen. Even then, every plan is based on assumptions and some of these will be challenged as plans meet with reality. Even project plans for similar projects, implemented multiple times, will differ due to the unique circumstances of different site stakeholders and the desire to apply improvements due to lessons learned.

5.1.1 Milestone Planning

To support the need for a realistic and achievable project plan that can adapt to the evolutionary nature of projects, a hierarchy of project planning tools is used (Figure 5.3).

Milestone plans set out the signposts of the project journey. This includes EEM stage gates, each of which has defined completion criteria to confirm that the desired outcome has been achieved before progressing to the next step. For example, do we have buy-in from all stakeholders to the *high-level design* (HLD) specification that will be used to develop the *invitation to tender* (ITT) document? Once past the stage gate we can then begin circulating it with confidence. These stage gate exit criteria provide a project quality assurance (QA) framework.

The quality plan extract shown in Figure 5.4 includes results paths indicating the relationship between activities and activity completion criteria. Work can start on any activity at any time, but an activity cannot be completed until the preceding activity has met the completion criteria. This provides a means of managing the iterative, evolutionary linkages between the people-, procedure-, and process-related developments across the EEM steps.

Milestone plans are supported by

- Activity schedules that define what has to be achieved to progress to the next milestone.
- Tasks that set out how the activities will be completed. These are also used to identify skill and resource levels.
- Briefing notes setting out guidelines for the aims, approaches, and outputs for activities/tasks (see Figure 5.5).

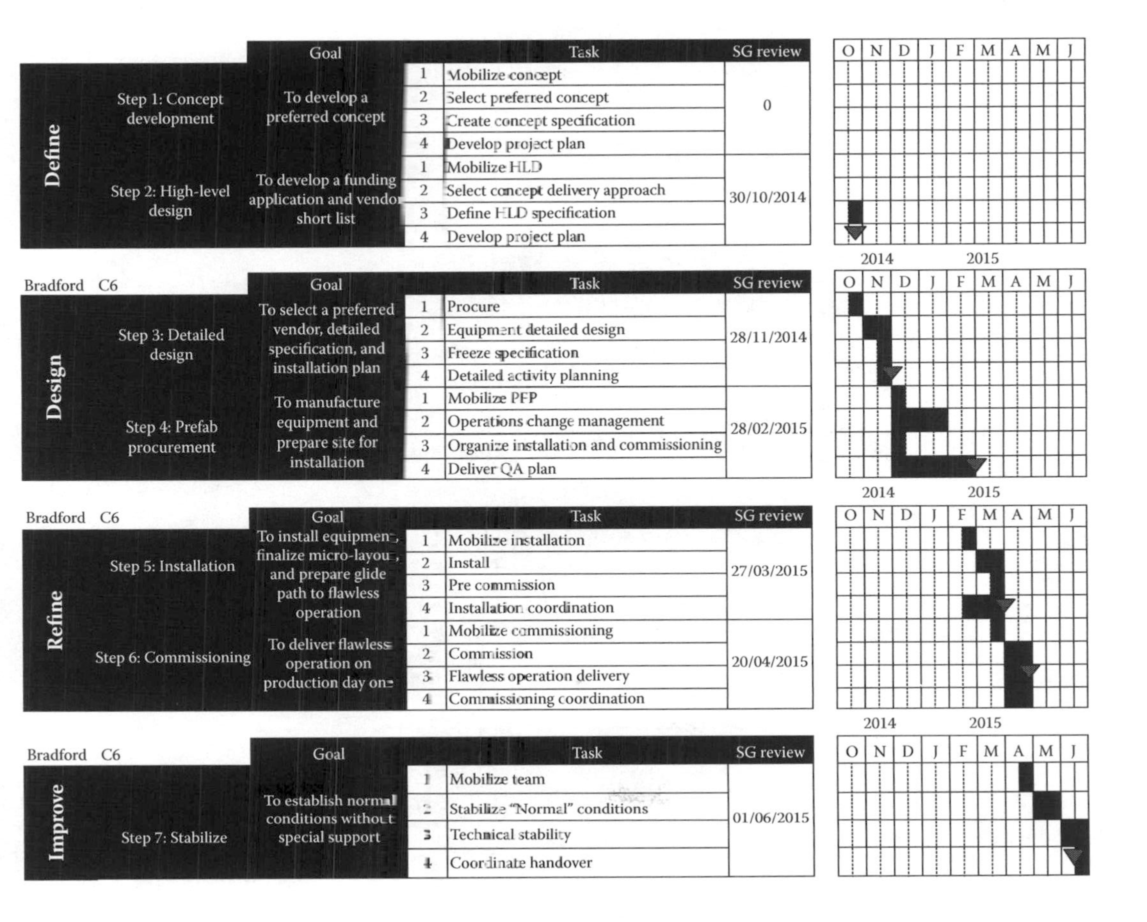

FIGURE 5.3
EEM milestone plan (example).

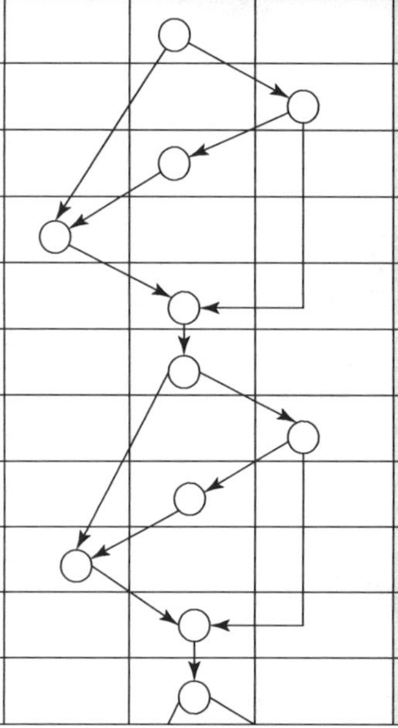

Quality plan

	Results paths				
No.	Procedure	Process	People	Milestone	Completion criteria
1				Start	
2				1.1 Mobilize concept	Clarify brief, identify people to be involved, and hold first team meeting
3				1.2 Select preferred concept	Objective testing and selection of preferred option
4				1.3 Create concept specification	DILO and design assessment
5				1.4 Develop project plan	Action map development
6				1.5 Concept stage gate review	Audit 1 and stage gate review
7				2.1 Mobilize HLD	Prep for objective testing of HLD options
8				2.2 Select concept delivery approach	Objective testing HLD
9				2.3 Define HLD specification	HLD module/VE review
10				2.4 Develop project plan step 2	Action mapping DD
11				2.5 HLD stage gate review	Audit 2 and stage gate review

FIGURE 5.4
EEM road map quality plan (example).

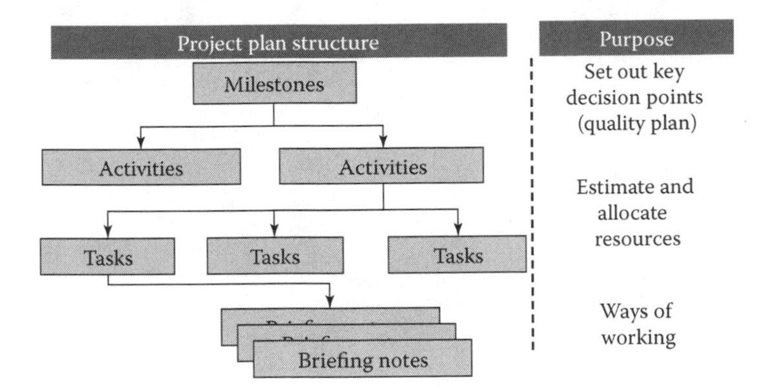

FIGURE 5.5
Critical path/slack time analysis.

In summary, milestone plans are used to

- Define the main signposts on the journey
- Aid the identification of skills and resource levels for each task
- Allow detailed planning to be left as late as possible
- Simplify progress measurement and reporting
- Provide a top-level project quality plan to confirm progress.

5.1.2 Understanding the Critical Path

When developing the project plan it is important to understand which tasks are on the critical path. Each activity schedule should contain sufficient information to support the identification of the earliest start time, the duration, and the latest finish time to support the assessment of the critical path. At the early stages of a project, this will generally be related to the completion of processes such as obtaining statutory permissions or long lead time items (Figure 5.6).

Later in the project this concerns the management of contractors, vendors, and internal resources. Depending on the size and complexity of the project, later in the project it may be worthwhile carrying out a formal review of the critical path—that is, those activities with zero slack between the earliest and latest start times.

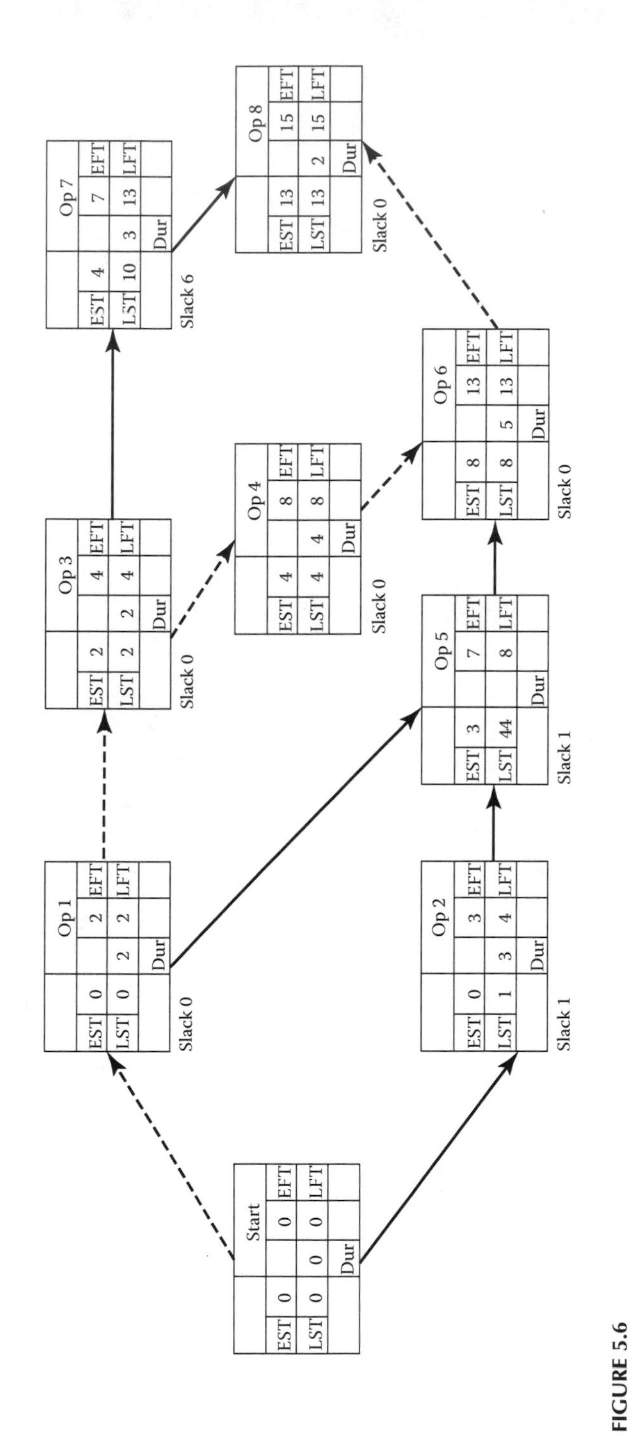

FIGURE 5.6
Developing a project plan.

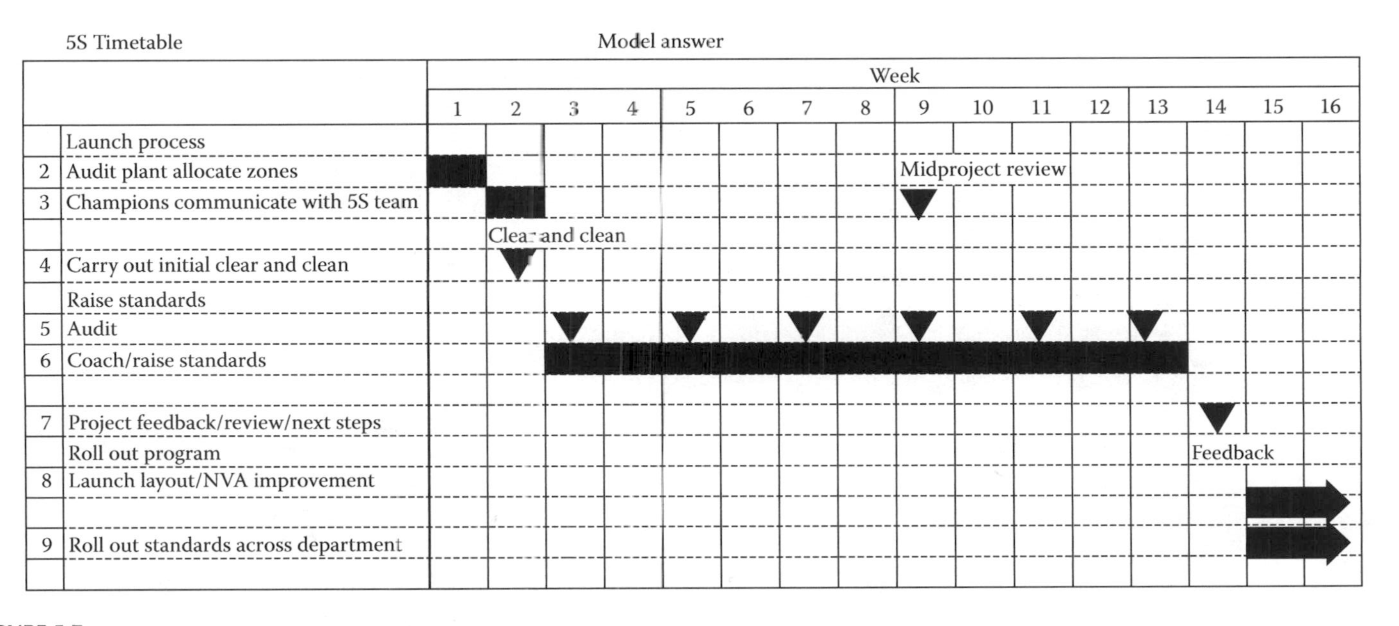

FIGURE 5.7
Gantt bar chart (example).

5.1.3 Communicating Project Time Lines

The popular Gantt bar chart format can then be used to communicate the sequence of tasks and project timings (Figure 5.7).

A common mistake with Gantt charts is to add too much detail too soon. Then, when changes to that schedule are needed, there is resistance to updating the plan because it will only change again and the effort taken to update the current plan will be wasted.

Although updating plans to allow for changes of dates is simpler with software tools such as Microsoft Project (other project management software is also available), these tools take time to learn to use, which means that updating them is not always straightforward.

As the EEM milestone plan sets out the activities needed for each step, good practice is to add detail to that to cover the current and next milestones. As the team approaches each stage gate they confirm the resources needed for the next step.

5.1.4 Work Scheduling

As mentioned previously, the purpose of planning is to set out a realistic and achievable program to coordinate the provision of resources where and when needed.

Scheduling concerns the detailed sequencing of tasks to deliver the program based on the resources available. The importance of efficiently scheduling installation and commissioning resources cannot be over emphasized. The definition of these tasks begins at the HLD step as part of the functional definition of design modules. This is where the input, process steps, and outputs are first defined.

As the project progresses to ITT and detailed design activities, technical requirements are finalized. This detail can then be used to define installation and commissioning work packets, which are then used to create the schedule of work. See commissioning programme evolution at Figure 5.8.

5.1.4.1 Visual Management of Work Packet Schedules

Visual schedule progress tracking is simpler if work packets for installation and commissioning are created using a standard time interval. The standard duration of work packets makes it easier to optimize plans and manage scarce resources. In addition, during work completion, delays or early completions are visible at a glance.

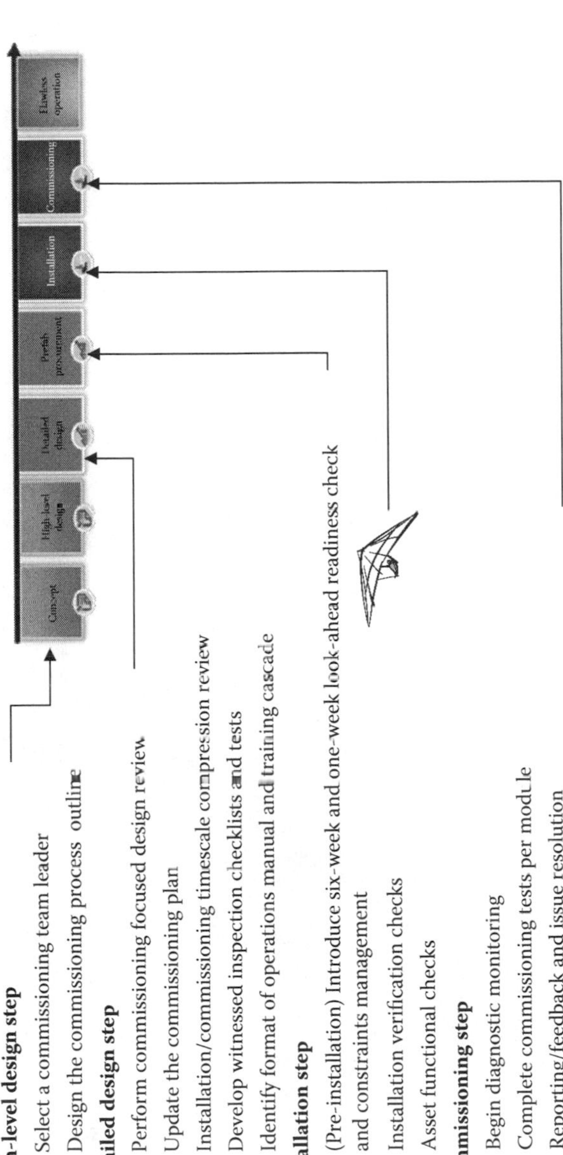

FIGURE 5.8
Commissioning program evolution.

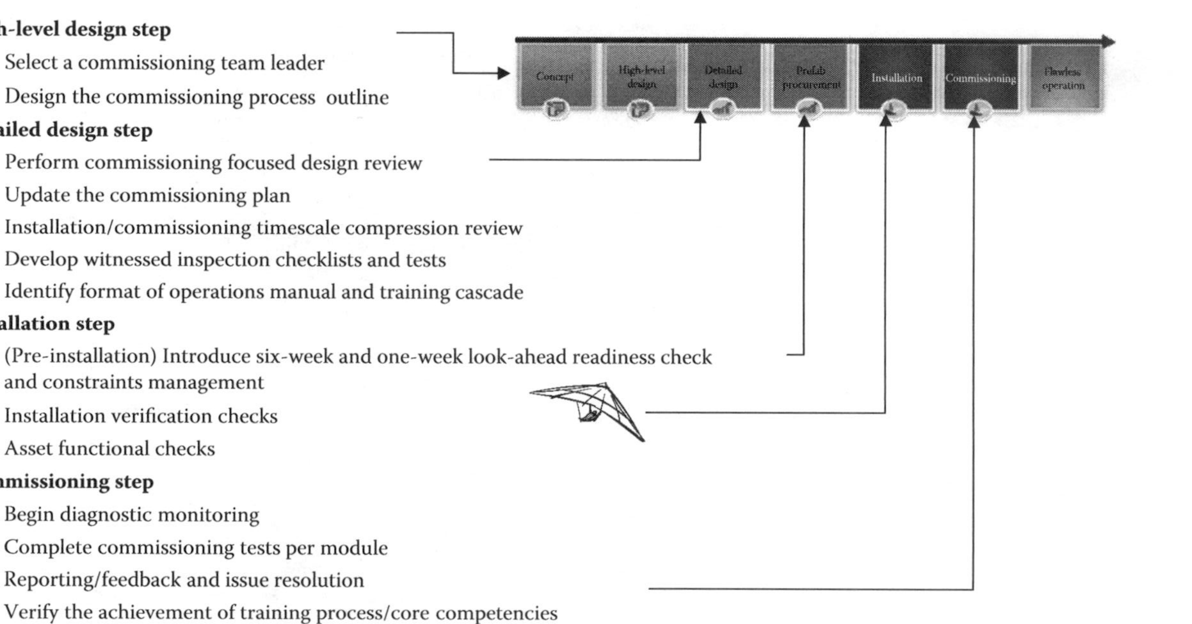

FIGURE 5.8
Commissioning program evolution.

Two hours is a useful time slot. Under this regime, if work is planned for a four-hour duration, the work packet definition includes midpoint completion criteria. At the midpoint, schedulers then get an early warning of delays and can provide additional manpower to help finish the task on time. Where the work involves modifications to existing equipment, an assessment of the need for additional unforeseen work is carried out in the first two-hour slot. That way the forward schedule can be amended to take this into account. The visual schedule makes it easier to understand the knock-on effects of any changes.

In the automotive industry, where the introduction of new models results in the regular rebuilding of production lines, the adoption of a two-hour scheduling work packet window delivers major gains in on-time project delivery. This facilitates the scheduling of critical resources and makes delays and early completions visible at a glance. Think of this as a schedule of four-hour blocks of activity with a midpoint measure of progress. The use of fixed time slots for work packages and visual progress reporting simplifies work load balancing, site resource management, progress reporting, and project quality assurance (Figure 5.9).

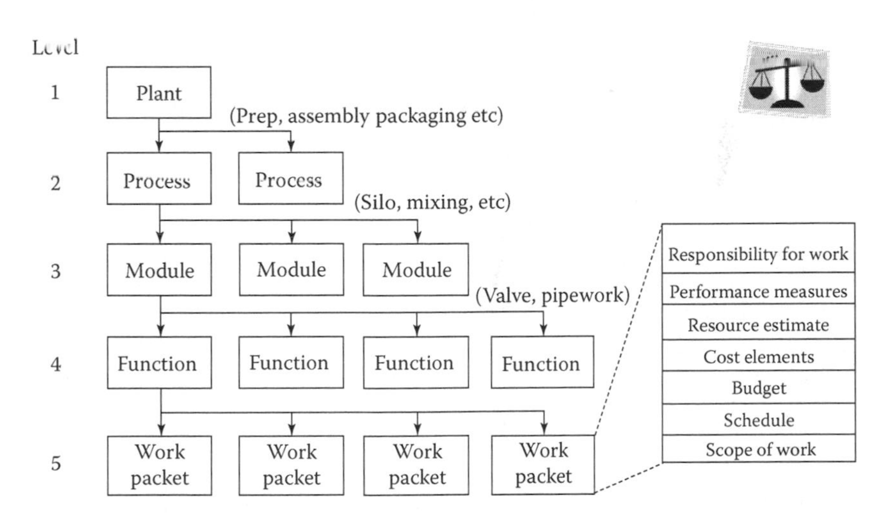

FIGURE 5.9
Work packet structuring.

Track the progress of installation planning against the following milestones:

- Work packet content readiness
    - Method statement
    - Power supply/scaffolding/lifting/transport
    - Completion criteria/inspection and testing
    - Risk assessment and mitigation plan development
    - Emergency procedure
- Health and safety assurance protocol
    - Design safety patrols, safety partners, and daily accident prevention activities
    - Arrange resources to carry out safety inspections of contractor equipment
    - Provide safety training/induction facilities
- Planning
    - six-week forward plan
    - one-week forward schedule
- Work release, reporting, review, and action protocols
    - Safety training
    - Safety briefing
    - Work permit completion (hot work, confined areas, work at height, etc.)
    - Assigned work area coordinator
        - Rehearse emergency drills
        - Audit compliance
    - Defined toolbox meeting schedule
- Work completion
    - Testing/inspection results
    - Feedback/lessons learned
    - Contractor compliance with plan

Table 5.1 sets out an example project work schedule. The planning protocol incorporates a six-week and one-week look ahead. The six-week look ahead is used to assure the quality of contractor plans and to ensure that there is sufficient time to procure additional resources or manage clashes if necessary. Vendor performance tracking includes the percentage of promises fulfilled, which can also be incentivized to reinforce good practices and as a lever against unrealistic promises. Ideally the six- and one-week look-ahead planning sessions are attended by all organizations that are involved with the work in that week, which provides opportunities for collaboration and clarification of who does what and when.

TABLE 5.1

Coordinating Installation and Commissioning Steps

Status	3–4 Months Ahead	6 Weeks Ahead	1 Week Ahead	On the Day	Day + 1	Week + 1
Execution						
Plan to do	Milestone plan					
Can do		Agree program, commit resources				
Will do			Confirm readiness			
Doing				Confirm status		
Done					Report completed	
Learned/approach refined			Confirm first-run studies	Confirm first-run studies	Trend percentage completion as planned percentage promises kept and actions to confirm glide path	
Supporting planning tasks	Develop workstreams, output steps, co-dependencies, and accountabilities	Choreograph detailed work packets and resources; confirm promises	Confirm readiness re prior work, materials, space, and preparation	Confirm progress as planned	Confirm/capture lessons learned	

5.2 MANAGING PEOPLE AND TEAMS

The success of a project depends as much on the project team culture and the team's relationships with the rest of the organization as it does on their knowhow. A project manager needs to be able to facilitate this network interaction to deliver the full project potential. They are part leader, part teacher, and in some cases part social worker.

This facilitation role concerns the interaction of three sets of project stakeholders:

- The EEM core team
- The stage gate team
- The rest of the organization

The relationships between them and accountabilities for each step are set out in Figure 5.10. (This is covered in more detail in Chapter 2.)

The scope of this facilitation role relates to the project process (how things are done) in addition to the project task (what is done). That concerns actions to

- Manage uncertainty, knowledge, and consensus
- Educate and raise understanding
- Set expectations and reinforce good practices
- Manage change and the communication of bad news
- Keep those involved in day-to-day operations connected with the project process and vice versa

Project managers should become familiar with the appropriate use of three facilitation styles. These are:

1. Directive
    - Sets objectives, steers the learning process.
2. Cooperative
    - Helps to structure team goals and priorities; supports decision-making processes.
3. Self-directed
    - The team has the freedom to select their own way; the facilitator creates the environment for self-learning and provides feedback/input as requested (Figure 5.11).

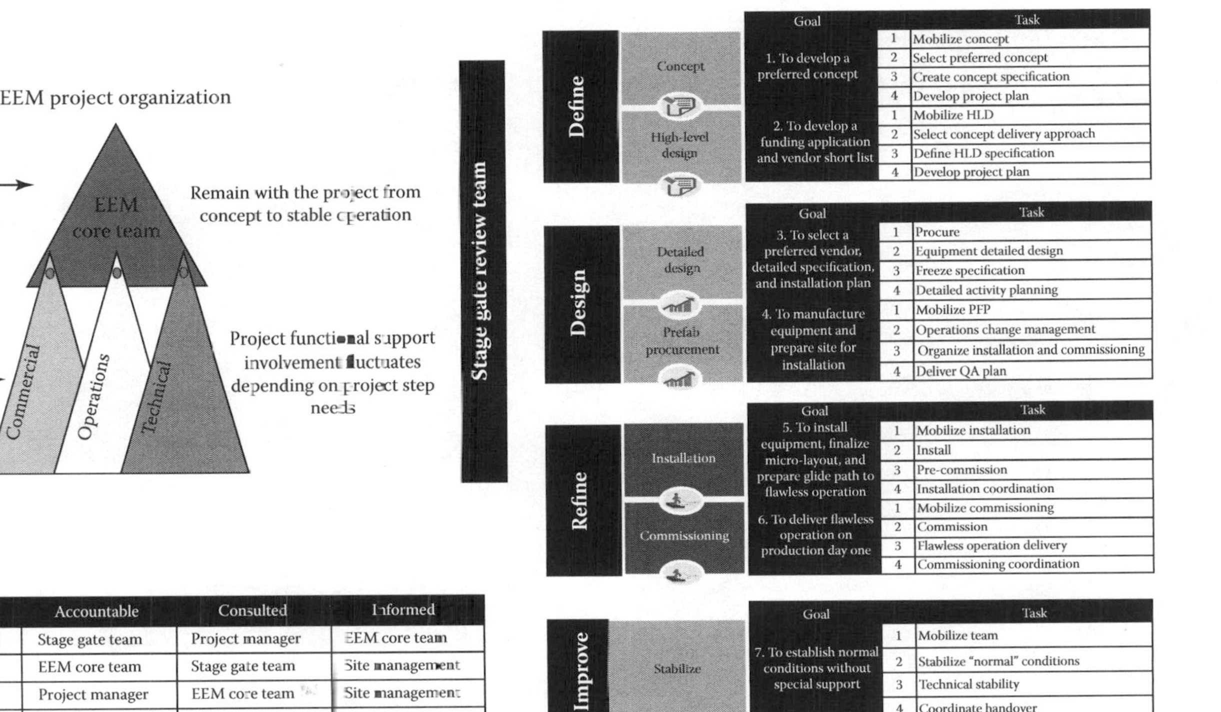

	Responsible	Accountable	Consulted	Informed
1	Site management	Stage gate team	Project manager	EEM core team
2	Project manager	EEM core team	Stage gate team	Site management
3	Stage gate team	Project manager	EEM core team	Site management
4	Stage gate team	Project manager	EEM core team	Site management

Goal		Task
1. To develop a preferred concept	1	Mobilize concept
	2	Select preferred concept
	3	Create concept specification
	4	Develop project plan
2. To develop a funding application and vendor short list	1	Mobilize HLD
	2	Select concept delivery approach
	3	Define HLD specification
	4	Develop project plan

Goal		Task
3. To select a preferred vendor, detailed specification, and installation plan	1	Procure
	2	Equipment detailed design
	3	Freeze specification
	4	Detailed activity planning
4. To manufacture equipment and prepare site for installation	1	Mobilize PFP
	2	Operations change management
	3	Organize installation and commissioning
	4	Deliver QA plan

Goal		Task
5. To install equipment, finalize micro-layout, and prepare glide path to flawless operation	1	Mobilize installation
	2	Install
	3	Pre-commission
	4	Installation coordination
6. To deliver flawless operation on production day one	1	Mobilize commissioning
	2	Commission
	3	Flawless operation delivery
	4	Commissioning coordination

Goal		Task
7. To establish normal conditions without special support	1	Mobilize team
	2	Stabilize "normal" conditions
	3	Technical stability
	4	Coordinate handover

FIGURE 5.10
EEM organization, milestones, and RACI chart.

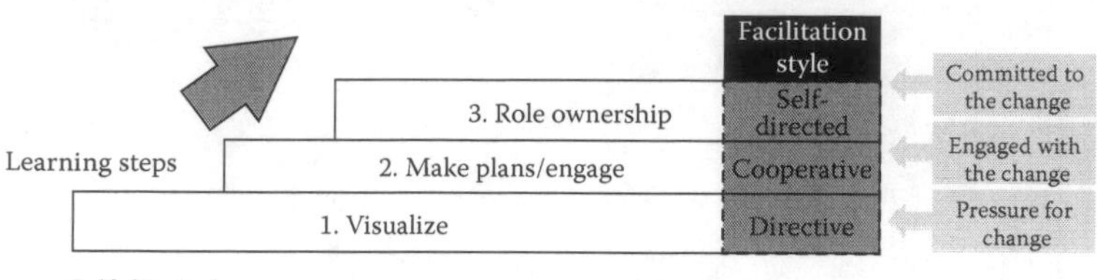

- Self-directed
  - Team has freedom to select their own way, facilitator creates the environment for self learning and provides feedback/input as requested.
- Cooperative
  - Helps to structure the teams goals and priorities supports decision-making process.
- Directive
  - Sets objectives, steers the learning process

Vary the facilitation style with the project stage ... and local leadership style

FIGURE 5.11
Project manager facilitation role.

The purpose of project management facilitation is to structure project teamwork tasks so that team behaviors progress toward self-directed competence, the planned outcome being the engagement of individual team members with a collective purpose.

The behavioral signposts that mark the steps of this transition are as follows:

1. Team members are able to understand the task and visualize what needs to be done.
2. Team members make plans to achieve the task.
3. Team members implement those plans.

The progression from point 1 to 2 in this list signpost 1 to signpost 2 is an important behavioral transition. Without this transition, it is not unusual for team members to wait for tasks to be allocated and then later complain that they had nothing to do. Their ability and willingness to make the transition will be influenced by the culture within the project site. In particular, front-line personnel involved in projects for the first time need help to move from a directive, defined production environment to one of uncertainty and discovery. The appropriate facilitation style will be to initially replicate this directive approach but allocate tasks to supports the transition from step 1 to 2.

To accelerate this progression, give new team members the opportunity to develop plans or define their role in the team from an outline brief. Once their plan has been developed they have made the behavioral transition from step 1 to 2. This sets the foundations of the psychological contract between the team member and the project manager. The project management role is then to help refine the plan and support the delivery of it.

Get the team engagement right and the project management role is transformed from one of policing to one of project leadership—much more effective and rewarding approach.

5.2.1 Creating the Collective Team Vision

Feedback from 6000 project team members* highlights the common pitfalls that impact on team performance and the following main contributory factors:

1. The quality of team member induction
    - Planning is inconsistent.
        - Patchy knowledge of project methodology.
        - Confusion over roles and responsibilities.
        - Authority levels are unclear.
        - Poor discipline/inconsistent project management practices.
    - Knowledge is hard to find.
        - Documentation management systems are inadequate or duplicated.
        - Access is restricted even to those on the project.
        - Cabinets are full of unused out-of-date documents.
2. Poor working relationships and team dynamics
    - Collaboration is limited.
        - The project team is dispersed in multiple locations.
        - Ad hoc networking but no formal process in place.
        - Knowledge sharing is on a one-to-one, ad hoc basis.
        - Decisions are made in isolation, poorly communicated, and of low quality.
    - Limited learning from experience
        - Lessons are not passed on from site team to project team.
        - Resources are not allocated to the task of capturing the lessons learned.
        - Designers don't use experience from past projects.
        - Mistakes are repeated.
        - Blame culture encourages the hiding of mistakes.

* Frank LaFasto and Carl Larson, *When Teams Work Best*, Sage, Thousand Oaks, CA, 2001.

5.2.1.1 Team Induction

An important lever in the creation of team culture is the team member induction process. Include the managers of team members released to the project in this. Some managers will see the release of personnel to the team as the end of their involvement in the project. In reality, they need to retain their management and leadership role during the project because

- The team members will have been selected because of their skills, knowledge, and experience; they will be working on behalf of their department. As such, they will be expected to confer with their colleagues and agree working assumptions, design guidelines, and technical standards. Their recommendations must be the recommendations of their department head.
- Team members will need to maintain close links with day-to-day operations to communicate progress and coordinate data capture, analysis, and documentation updates at each step.

For larger projects it is worth setting out a formal team induction pack covering

- The team process
    - Induction: Team vision, team process, team values
- Standards of organization
    - Project management documents
    - Personal administration
- Document management and version control
    - Where to find information, how to update

5.2.1.2 Team Dynamics

It is important to put in place the foundations for an effective team relationship during the early steps of the project by working with the team to develop

- A vision statement describing the outcome required and measurable goals
- Clear accountabilities and using audits/coaching
- Quality plan milestones with defined exit criteria
- Channels for feedback, including bad news

The definition of a clear project vision, as discussed in Chapter 2, helps teams to deal with uncertainty using a systematic decision approach. Work on the vision also forms part of the action mapping process, which helps to align priorities and integrate parallel work streams.

Team dynamics can be influenced by team member behavioral profiles. Where this is not known it can be useful to get team members to complete a profile so that they understand each other's strengths and weaknesses.

5.2.1.3 Collaborating with Vendors

Vendors are an important part of the project team. The most effective approach to working with them transcends the traditional vendor working relationships illustrated in Figure 5.12.

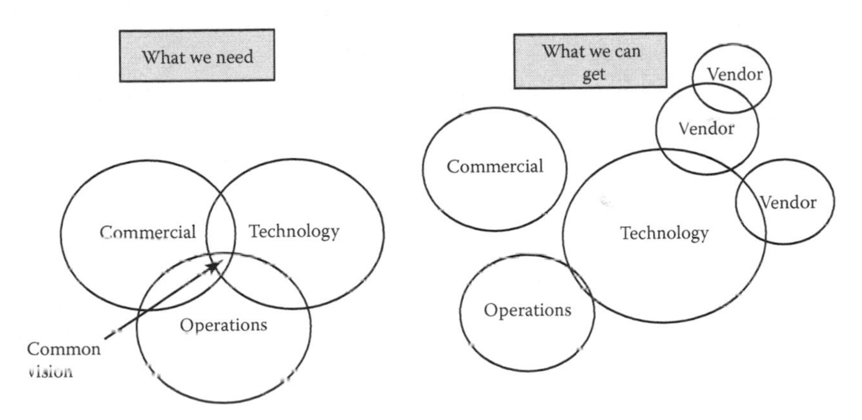

FIGURE 5.12
Defining proactive working relationships.

The foundations of vendor working relationships are laid before the contract is awarded. Work with vendors prior to selection to raise their understanding of what is needed; learn about what is available and the relative value of offerings. This provides useful insight into vendor skills and your own knowledge gaps. It will also help the team to define the criteria for selection, such as

- Commercial status, size, reputation
- Delivery capability
- Process capability
- Technical status
- Supplier culture

- Financials/prices
- Ability to provide support

The following actions will support the development of a productive collaborative working relationship.

- Share information and priorities.
  - Be clear about what you need and what is important.
- Always get competitive bids but allow vendors to improve your approach.
  - Be clear about what "good" looks like, but let the vendor use their expertise.
- Build partnerships for the long term.
  - Collaboration is more effective than hands-off transactional relationships.
- Seek to understand your vendor's business too.
  - Know what your business means to the vendor and what they could be able to negotiate on.
- Negotiate to achieve a win–win agreement.
  - Don't try to offset risks that are outside of the control of the vendor.
- Focus on value for money.
  - Don't focus only on the lowest cost.

Underpin the working relationship by working with the vendor team to agree a project charter that emphasizes

- A collaborative rather than arm's-length approach
- A willingness to learn rather than an autocratic outlook
- Negotiating on value rather than lowest cost

5.2.2 Managing Communication Processes

Good communication processes will stimulate stakeholder interest and engagement as well as reinforce team values and maintain team morale. This includes

- Daily/weekly progress, delays, resource usage versus budget for each work package, progress against quality plan milestones
  - Record the resources used and potential for improvement.
  - Record safety patrols, incidents, and actions taken.

    - Work completion/start-up report, planned versus actual problems and opportunities.
    - Preplan for the next step.
    - Designate a single point of contact to provide regular, honest one-page progress reports.
- Structured, formal problem-solving
    - Defect management, query resolution, defect recording, next steps.
    - Be prepared to stop and take stock.
    - Documented problem-solving steps, conclusions, and lessons learned.
- Systems for assuring progress quality and controlling changes
    - Test/audit results
    - Change approval and documentation update

Creating a visual project workspace or "war room" encourages an open system of reporting as an aid to communication and stakeholder engagement. Typical facilities in the project war room include

- Computer and projector linked to an internal network
- Project plans
- Whiteboard/flip charts
- Wall charts showing project vision and progress to date
- Outstanding issues map
- Project plans

Research into the use of visual management by manufacturing organizations* reveals that those organizations that use visual management well achieve a higher level of problem resolution. This is because visual management makes clear the need to make decisions; as a result, those in the area act on that information rather than wait to be told what to do next.

The same is true of the visual reporting of project issues and project progress. Use reporting formats that lend themselves to visual reporting, such as the status wheel shown in Figure 5.13. Here, each quarter of the status wheel indicates the completion of the sequential tasks of *value engineering*, *residual risk review*, *best-practice routine development*, and

---

* Michel Grief, *The Visual Factory*, Productivity Press, Portland, OR, 1989.

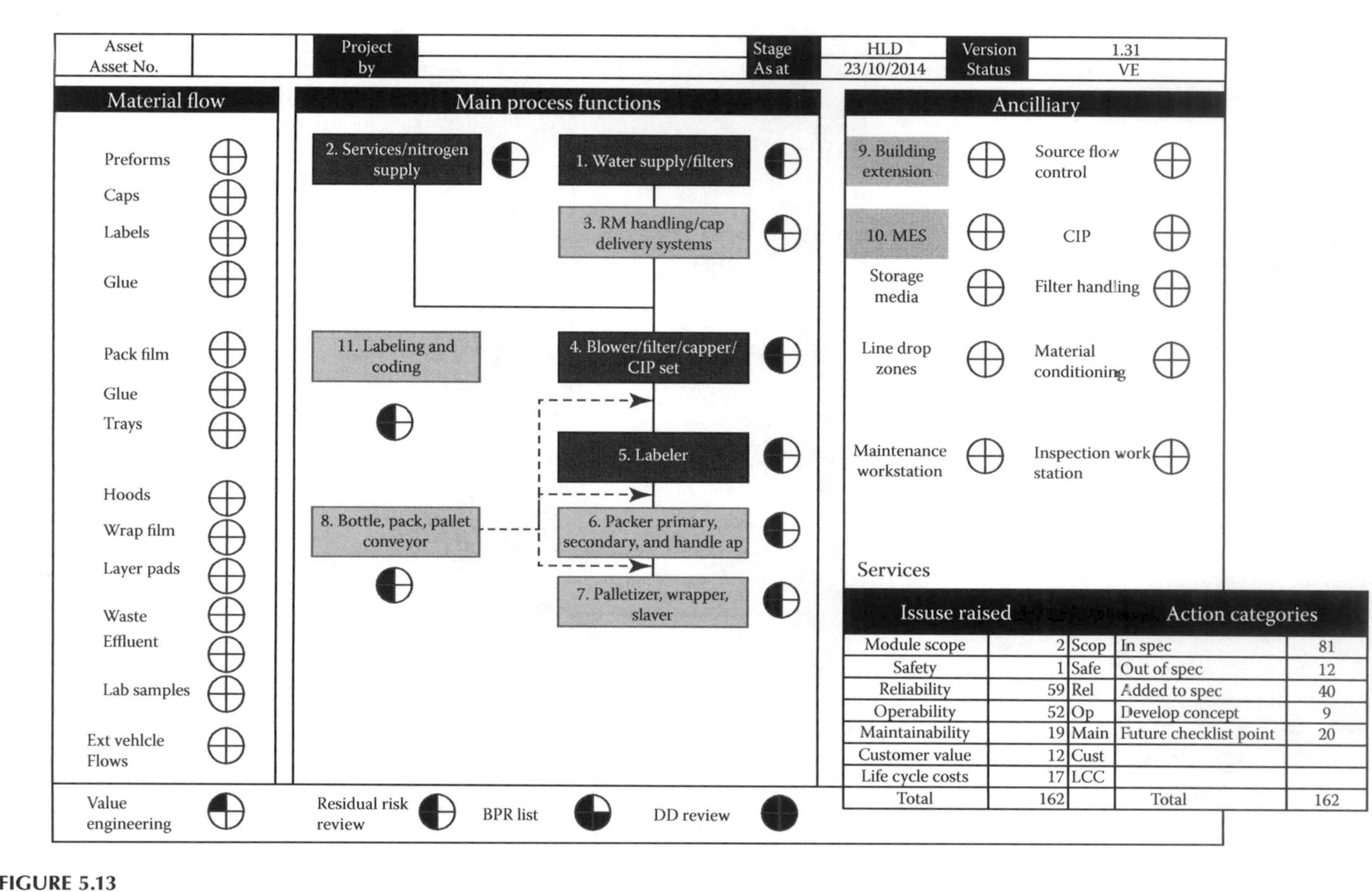

FIGURE 5.13
Module review status wheel reporting.

detailed design for each module. Also shown are the cross-module issues raised. The status shown here shows that all modules have completed the value engineering and residual risk reviews. The preparation of best-practice routines prior to detailed design has yet to be completed. As the next project step is HLD, work is on target.

The project environment is one that needs to be able to deal with uncertainty. People in that environment need to absorb and deal with issues as they arise. This is in comparison with the more structured operational environment where there is more certainty and practices can be fine-tuned. It still needs to be adaptable to changes in demand, but people in that environment are supported by systems and procedures to assure the consistency of supply to customers. The project environment has fuzzy, almost sponge-like boundaries compared with the operational environment with its clean, well-defined boundaries. There is a significantly different working approach in each of these that presents a challenge to those joining the project and those working closely to it (hence the importance of team induction and the repetition of messages to assure understanding) (Figure 5.14).

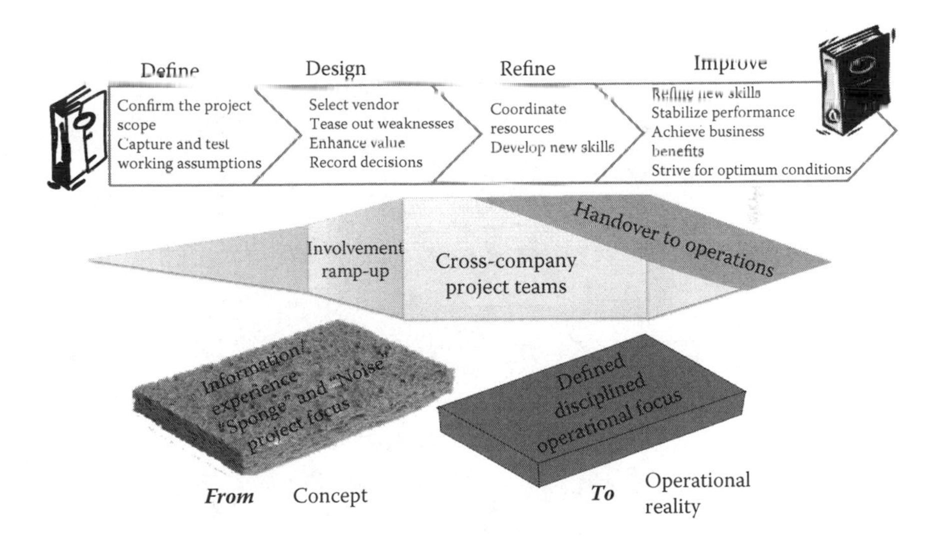

FIGURE 5.14
The project team challenge.

5.2.3 Project Administration

Project administration provides the conduit for information flows, document management, and project resource management. This includes compliance with statutory legislation such as the Construction Design and Management (CDM) Regulations 2015 and the Provision and Use of Work Equipment Regulations (PUWER) 1998. These are discussed in more detail in Chapter 6, "Project Governance."

Formal systems include

- Project logistics processes
  - Contract management
  - Planning
  - Scheduling
  - Execution
- Rooms/storage (floors, shelves, rooms)
  - Workplace layouts
  - Office equipment (photocopier, facsimile, telephone)
  - Computer access and storage locations
  - Equipment and tooling
  - Fixtures and fittings (folders, files, lockers)
  - Stationery
- Document locations
  - Notices, standards, filing, reports, record keeping
  - Reference systems/labeling

The smooth running of these processes has a major impact on teamwork. Figure 5.15 is an extract from the lessons learned review for a

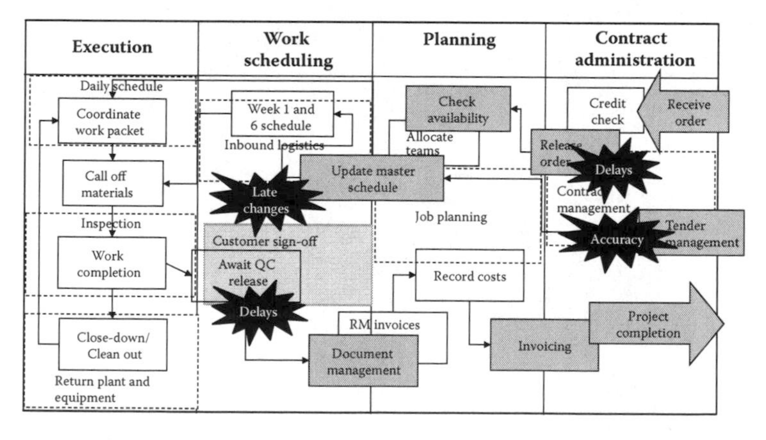

FIGURE 5.15
Underpinning project processes.

project that suffered from delays due to project administration problems. Time spent implementing and debugging project administration is time saved.

5.2.3.1 Personal Systems and Procedures

Personal administration procedures also have a significant impact on project progress. These include

- Communication
    - E-mail
    - Contact database
    - Web portals
- Time/task management
    - Calendar
    - Appointments
    - To-do list
    - Notebook/daily log
    - Learning log
    - Acting on issues as they arise (the one-touch concept)
- Procedures
    - Daily, weekly, monthly routines/standard work
    - Document management
- Document management
    - Information folders
        - Data
        - Analysis
        - Key documents
        - Presentations and reports
    - Templates
        - Meeting minutes
        - Progress reporting
        - Cost control
        - Plans and resources

Team members who are new to projects may need help in developing these personal systems.

5.3 MANAGING RISK

5.3.1 The Purpose of Risk Management

As introduced in Chapter 4, the purpose of the design risk assessment is to reveal safety hazards and minimize the risk of accidents during installation and commissioning work and during equipment operation.

This introduced the risk mitigation hierarchy—eliminate, reduce, isolate, control, protect, and discipline (ERICPD) developed to manage technology and operational risks. The sound logic of the hierarchy can also be applied to the management of other project risks—for example, when considering financial or market risks.

Different levels of risk are considered at each project governance phase.

- Risks within the *define* phase
    - Financial, market, customer, and volume risks
    - Scope
    - Process capability
- Risks within the *design* phase
    - Design intrinsic reliability and safety
    - Vendor disputes/delays
    - Cost/time overruns
- Risks within the *refine* phase:
    - Installation risks
    - Commissioning/validation failures
    - Health and safety
    - Human error
    - Contamination (Figure 5.16)

5.3.2 Surfacing Hazards

EEM design guidelines/standards for safety should include definitions for acceptable standards of common risks such as access, ergonomics, utility hazards, equipment hazards, fall hazards, isolation, line of sight, material flow, micro-hazards, product safety/QA, and statutory compliance. Add to this indigenous hazards such as dust control in wood yards or flour mills. These design guidelines are applied at the concept, high-level, and detailed design steps to remove the causes of potential hazards where possible. The use of EEM guidelines will then be sufficient to support the specification of ITT documents.

1. During define steps
   - Rigorous assessment of preferred option
   - Actions to deliver improvement priorities/targets
2. During design steps
   - Confirm detailed design and plan for installation step
   - Develop provisional best practice to explore cause/effect mechanisms
   - Focus on problem prevention starting with critical modules
3. Near end of design steps
   - Confirm installation/ pre-commissioning and plan commissioning activity
4. During refine steps
   - Confirm commissioning and prepare for run up to day-one operation
   - Refine best practice, visual management and training processes
   - Focus on delivery of flawless operation goals

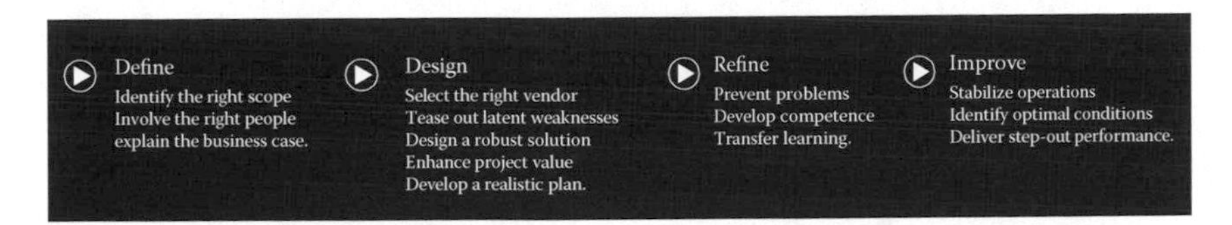

FIGURE 5.16
Surfacing hazards and managing risk.

The risk assessment process in the following section is designed to reveal the *residual hazards* of the selected vendor option as part of the detailed design process. This includes dynamic and consequential risks—that is, what could happen when things go wrong? What generic safety themes do we need to explore in more detail?

5.3.3 Registering Risks

Use the outputs from the module review as the starting point for recording potential hazards that could occur based on the way the equipment will be used. Avoid the mistake of assessing risks without a clear method statement. This can result in costly safety features that provide poor safety. For example, the specification of guarding, which does not consider routine access needs, can increase rather than reduce the risk of injury. Furthermore, the use of a method statement could lead to the refinement of working methods that eliminate hazards so do not need guarding.

During the detailed design workshops, update the module review working methods to account for specific isolation routines. Also confirm/refine safe systems of work to control and avoid creating the conditions that could result in a hazard. Finally, consider what personal protection could be needed to protect against residual risk and ensure its application by taking steps to make safe working practices part of the routine.

Maintain a register of hazards identified and make an assessment of the consequences of each hazard in terms of severity and the probability of that risk occurring.

A common approach involves the identification of potential hazard impacts and the probability of occurrence using a numerical scale (1 = low, 3 = high). These assessments are then multiplied together to categorize risks as low risk (1–3), medium risk (4–6), and high risk (7–9) (Table 5.2).

TABLE 5.2

Risk Assessment Matrix

		Impact	
		High	**Low**
Probability	**High**	7–9	4–6
	Low	4–6	1–3

The risk register example in Figure 5.17 emphasizes the level of risk and indexes entries by assigned risk category.

Low risk scores are generally considered acceptable. Medium and high risk scores are subject to a further review to assess the ease of detectability. In the following example, the risk scores are multiplied by the ease of detection scores (3 = difficult, 1 = easy) to further define those areas that need the most attention (Table 5.3).

5.3.3.1 Highly Regulated Industries

In the case of highly regulated assets such as those covered by Control of Major Accident Hazard (COMAH), Integrated Pollution Control (IPC), and other regulatory requirements, additional industry-specific risk assessments such as *hazard and operability* (HAZOP) or *hazard studies* will be needed. This process should only be undertaken with the support of a specialist. The HAZOP process developed by ICI for chemical plants involves the uses of six different levels of hazard study. These align roughly with the six EEM steps. Make sure that the HAZOP agenda used reflects the needs of the EEM step.

5.3.4 Implementing Risk Mitigation

After the detailed design process, risk management involves making sure that risk mitigation plans are implemented. The tracking of risk mitigation plans is carried out through the witnessed inspection program.

No	Area	Risk	Owner	Outcome if no action	Mitigating action	Rev forum	Next review date/ closed	Accept/ mitigate	Prob	Impact	Risk
1	Molding	Molding plant order needs to be placed before final recipe sign off.	JT	Tech Spec does not contain specific product information and contract guarantees/ warranty is invalid	Vendor confident that if final masses are in the range that has been tested then the plant will produce to the standards/speeds required.	Steering	10/04 20xx	Accept	1	2	2
2	Production prep	Availability of services/ site/individual capacity	IAK	Additional costs/delays to upgrade infrastructure	Initial checks carried out and look OK	Enginee-ring	Closed	Accept	1	2	2
3	Cooking	Potential burn on to new dissolver at high temperature due to undissolved lumps of whey powder	JT	Possible that cannot increase line rate with dissolver also potential for foaming	Trial required to change order of ingredients to bulk up level.	Enginee-ring	11/05/ 20xx	Mitigate	3	3	9
4	Project	New process engineer for plant not yet identified or in place	PH	Potential issues in handover of plant and transfer of understanding-required ideally during commissioning at least	Require identification of plant process engineer to replace grads	Steering	11/05/ 20xx	Mitigate	3	3	9

FIGURE 5.17
Risk register (example).

TABLE 5.3

Risk Assessment Step 2: Ease of Detection

		Ease of Detection	
		Difficult	**Easy**
Risk assessment	**High risk**	19–27	12–18
	Medium risk	12–18	4–11

The witnessed inspection program uses two types of documents:

- Checklists of features to be provided
- Test plans

Checklist and test development begins at HLD as part of the module review process. This includes the creation of witnessed inspection categories for each design module, such as those set out in the following table. Some categories, such as construction, can utilize standard checklists, but these should be scrutinized to be sure that they are relevant. As mentioned previously, long checklists on their own are no guarantee of success. Checklists are an aid to thorough inspection by a trained/experienced observer (Table 5.4).

TABLE 5.4

Example of Witnessed Inspection Plan Checklist Categories

Witnessed Inspection Categories: Module Cooker		
Cat.	**Title**	**Summary**
1	Construction	Confirm that correct build/component/wiring standards have been used.
2	Safety	Conforms to legal and company safety standards.
3	User interface	Screen structure plus authorization levels.
4	Functionality	Check everything works. Is it what we expect prior to dispatch to site?
5	Efficiency	Confirm the asset target performance can be achieved.
6	Effectiveness	Confirm the asset design meets EEM standards.
7	Error handling/recovery	Confirm understanding of how it works and how to control the process.
8	Thermal testing	Confirm that the thermal profile meets requirements.
9	Documents/software	Confirm version and status of software, drawings, manuals, reference and training aids.

Test plans and checklists are confirmed and refined as part of the detailed design process and include the following tasks:

- Review vendor quality plans to confirm inspection criteria for
    - The manufacture of equipment prior to installation on-site
    - Installation plans, QA milestones, and inspection criteria during installation
- Create and agree test plans (Figure 5.18) and checklist points to confirm the achievement of quality plan exit criteria at key milestones, including
    - An equipment manufacture progress review
    - A factory acceptance test (FAT)
    - Building construction
    - Installation
    - Commissioning

Use witnessed inspection testing to provide evidence of progress toward flawless operation. This includes

- Equipment
    - Condition
    - Confirmation of progression past infant-mortality failure modes
    - Functions, features, and performance levels achieved
- Asset care regime
    - Start-up checks, running checks, servicing standard practices in place
    - Operating best-practice standardization of start-up, steady-state, close-down/clean-out routines
- Competency
    - Core competencies in place
    - Intermediate and specialist support in place
    - Skills transfer/competency development road map in place

Some tasks should only be begun following the successful completion of the relevant witnessed inspection test. This includes activities to

- Refine operations and maintenance working methods as part of commissioning
- Finalize visual management specifications and micro-level workplace layouts
- Complete training and skill development cascades

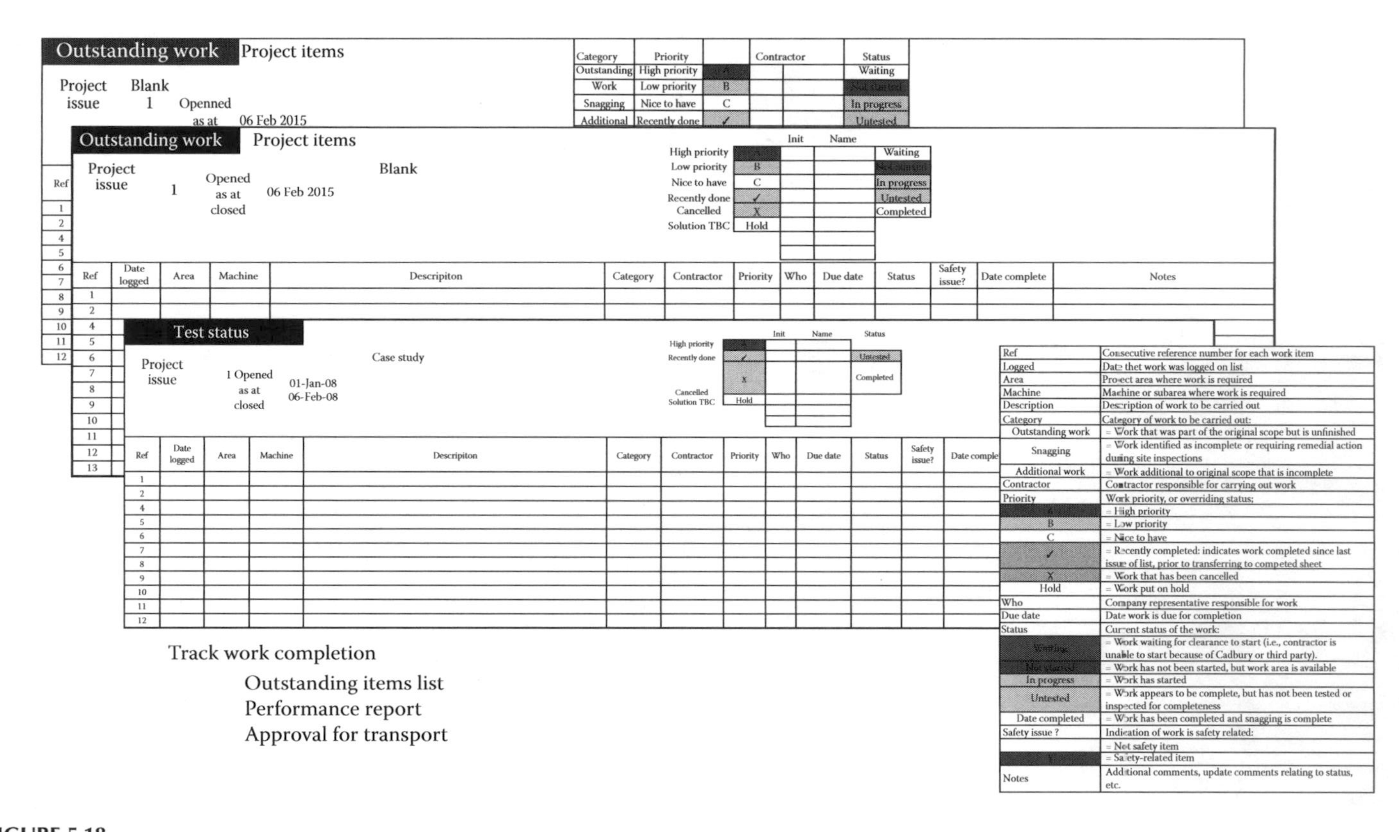

FIGURE 5.18
Witnessed inspection test records.

Figure 5.19 summarizes the steps of the witnessed inspection program plus roles and responsibilities from installation onward.

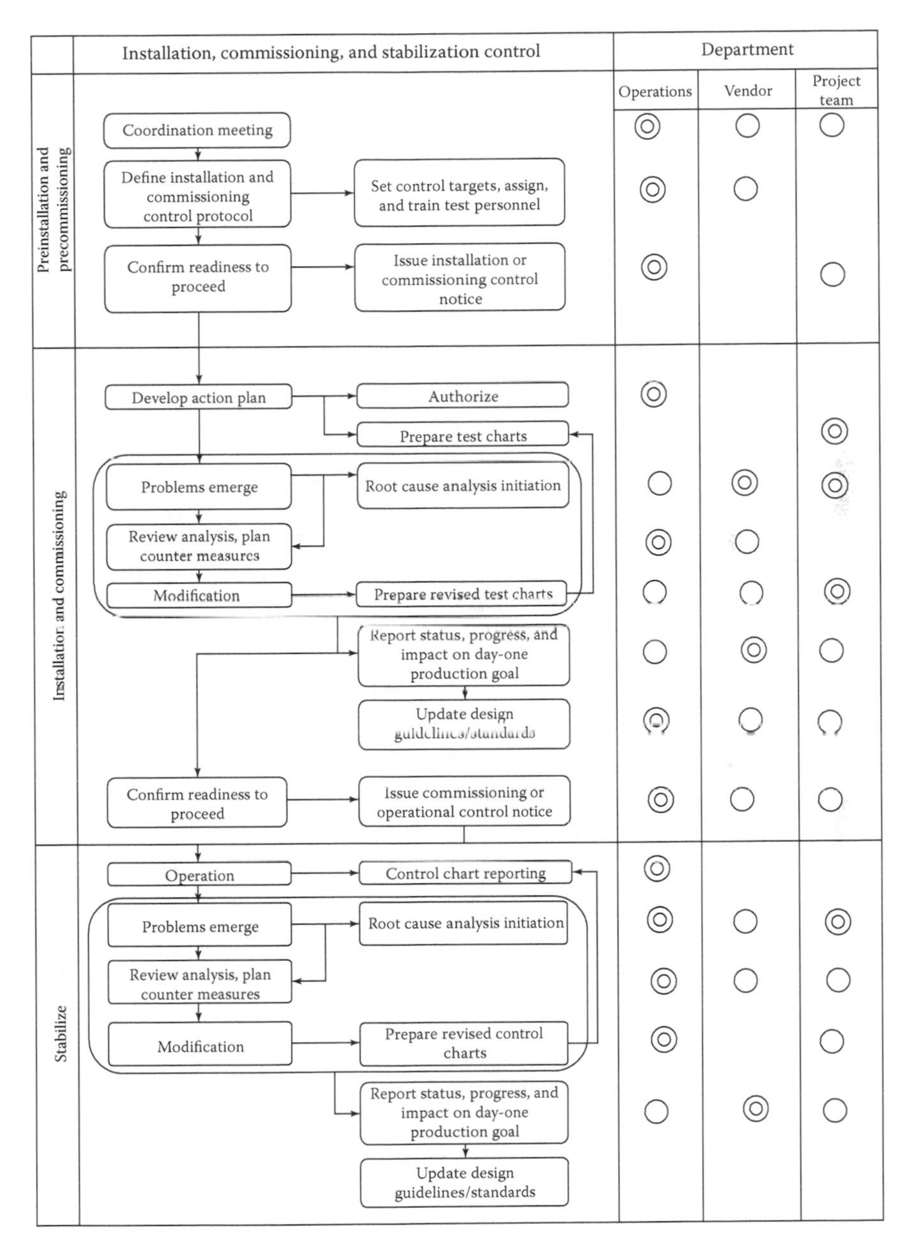

FIGURE 5.19
Managing the glide path to flawless operation.

5.4 DEFINE

The project and risk management process during the concept and HLD steps is as follows:

- Develop a forward program for EEM steps and align with the timings of other projects and events, such as planned production outages.
- Develop an outline specification and justification for progression to the HLD step.
- Define a quality plan, project organization, timing, and critical path.
- Develop a task list and timetable.
- Define the approach and resources to be used during the detailed design workshops.
- Create a project justification and complete the capital approval submission process.
- Create a risk register.
  - Identify at which EEM step they need to be dealt with.
  - Confirm the resolution of risks at each step.
  - Confirm with the team the presence of red and amber flag issues prior to the stage gate review.

5.4.1 Project Leader Facilitation Guide: Concept

1. Get the team together and review the concept.
2. Give yourself some time to discuss the brief with stakeholders and get the right people assigned to the project.
3. Work with the team to generate a long/short list of viable options and select a preferred approach.
4. Create a concept specification: fine-tune and refine to get the best possible value from the project. Identify the likely order of costs.
5. Develop a milestone plan with the team and assess potential roadblocks and opportunities. Develop the business case.
6. Stage gate review 1: Take the opportunity to test/challenge your logic with others who are impacted by the project but have not been fully part of the decision-making. Be prepared to stick to your guns. What you are looking for are resources to work up a more detailed justification.
7. Use this stage gate as a launchpad to the HLD step.

5.4.2 Project Leader Facilitation Guide: High-Level Design

1. Bring others up to speed. Discuss the preferred concept and agree on the following:
   a. What is the impact on current operations?
   b. What problems can we solve as part of the project?
   c. How will we get the best value from the project intervention?
   d. What performance targets should we set?
2. Split the project up into functional design modules to structure the design, manufacture, installation, and commissioning activities
3. Define relevant *design guidelines* and convert these into witnessed inspection categories to support project QA through to the commissioning and site acceptance tests.
4. Define the vendor selection process and timetable. What questions should we ask to assure they understand our needs? Begin to develop an draft ITT to support the process. What are the outline costs? Develop the base-case life cycle cost (LCC) model. How sensitive is this to volume/likely changes?
5. What are the hazards/risks associated with the design?
6. What are the main project quality milestones, resource needs, and timings? Are there any long-term purchasing items? What benefits can we expect from the project? Use the six EEM design goals to ensure the scope adds as much value as possible. Define the post-day one ramp-up program.
7. Submit the funding application.
8. Treat stage gate review as a launch pad for detailed design.

5.5 DESIGN

The tasks that are part of the project and risk management process during the detailed design and prefab procurement steps are as follows:

- Develop/refine the milestones, task list, and project governance timetable.
- Communicate the detailed design (DD) timetable.
- Risk assess the DD specification definition, transition, installation, and commissioning plans.
- Confirm the business justification and delivery quality plan.

- Communicate the detailed design and instigate change control. As part of this, define the document management coding, version control, author, editor, and access rights.
- Detailed activity planning and work packet development: Plan, organize, and control the detailed installation and commissioning plans with vendor/contractors.
- Define the witnessed inspection/QA program.
- Manage the construction phase, including witnessed inspection and problem prevention activities.
- Carry out the witnessed inspection program.

5.5.1 Project Leader Facilitation Guide: Detailed Design

1. Mobilize the team. This will include more functions, so spend time getting buy-in to the HLD. Ensure that there is buy-in to the vendor selection process. Collate ITT documents, including checklist points from HLD. Confirm the vendor's understanding of what "good" looks like and their ability to deliver as part of the vendor ITT response. Brief those negotiating contractual terms with vendors on the potential value of features that support the six EEM goals. Ensure that they are aware of the potential impact on LCCs so that these are not lost as part of price negotiations. (This is covered in more detail in Chapter 6.)
2. Organize vendor induction and visioning activity and detailed design workshops. During detailed design workshop 1 to ensure that the vendor design team understands the EEM Goals and guidelines that will be used to assess and sign off the final design. Explain 10% review the ways of working using a DILO approach to confirm that design goals, guidelines, and technical standards are understood. Agree dates for a follow up workshop(s) so that they can set out in detail how each module meets the EEM design goals and guidelines agreed as part of the vendor selection process. Following this, assess residual risks and firm up operational changes. Use LCC analysis and value engineering to refine how functionality will be delivered. Confirm problem and defect prevention measures.
3. Freeze the specification and instigate change management controls following
    - Risk assessment to confirm module risks mitigation tactics
    - Definition of organizational change management road map

- Confirmation of LCC forecast/improvement targets and how they will be met

4. Project management of the technical and organizational change road map, including the identification of tasks to assure the delivery of flawless operation.
5. Use stage gate review as a launch pad for the next step.

5.5.2 Project Leader Facilitation Guide: Prefab Procurement

1. Begin the journey to mastering new competencies, including the transition from the current/as-is to the new/to-be organization
2. Define core, intermediate, and specialist skills and design the skill development route. Make use of the *plan*, *do*, *check*, and *act* (PDCA) process to trial/refine working methods before finalizing the approach. Use modeling or simulation to support problem prevention and defect prevention activities at this stage. This will lead to the definition of visual management aids and the design of guarding/work locations/steps/floor markings, and so on.
   - This step should include visits to vendor reference sites to learn about the operational realities and test ideas for working methods.
   - Set up clear lines of communication between vendors and the EEM core team. Be prepared to organize at least two visits to carry out witnessed inspection prior to the predelivery inspection.
3. Work closely with vendors to agree installation and commissioning plans and complete risk assessment and project QA assessment. Use vendor installation plans, but expect them to demonstrate how they will manage the quality of their work.
4. Prepare for installation, including communication planning, site clearance, and allocating accountabilities for arrangements during the installation and commissioning stages. Develop installation and precommissioning communications packs.
5. Develop installation work plans for all internal and external activities. Aim to create balanced work packets of a fixed duration (e.g., two hours) to make it easy to use visual management to monitor progress and the coordination of installation and commissioning skills/resources. Use network analysis to define the critical path.
6. Carry out a readiness review well before the installation start date to be sure that all parties are prepared. Use six- and one-week look-ahead planning processes throughout installation and commissioning.

5.6 REFINE AND IMPROVE

The tasks that are part of the project and risk management process during the detailed installation and commission steps are as follows:

- Installation coordination and safety management.
- Use the witnessed inspection process to manage the glide path to flawless operation.
- Commissioning coordination.
- Assess day-one readiness.
- Defect detection and the collation of a list of outstanding items.
- Site acceptance tests confirm the performance meets the expected witnessed inspection test results.

5.6.1 Project Leader Facilitation Guide: Installation

1. Mobilize installation and induct new members, including training in safety protocols and the testing of safety routines, reporting, communications processes, and visual coordination schedules.
2. Coordinate witnessed inspection testing to confirm functionality. Conduct a formal review of issues and solutions; refine micro-layout issues and best-practice routines relating to movement and access.
3. Precommissioning and handover to commissioning team and risk assessment to identify readiness/flawless operation glide path.
4. Coordination to assure installation plan achievement and the capture of lessons learned.
5. Confirm technical stability and readiness to deliver flawless operation.

5.6.2 Project Leader Facilitation Guide: Commissioning

1. Mobilize commissioning team members. Track the glide path to flawless operations. Be prepared to stop, take stock, and refine the day-one date or increase resources as necessary.
2. Continue with issue capture and resolution activities. Feedback test results and update design books/documentation as necessary.

3. Super-users carry out training to support the development of core competencies and refine intermediate and specialist skill development routes.
4. Complete the witnessed inspection transfer and acceptance protocol for each module.
5. Sign off day-one readiness/validate the process prior to day-one production.

5.6.3 Project Leader Facilitation Guide: Stabilizing

1. Mobilize the operations team.
2. Sign off core competencies and the development of intermediate and specialist skills.
3. Assure technical stability (zero breakdowns) and contamination control. Define how to secure optimum conditions. Update manuals and technical documentation.
4. Coordinate handover, confirming plan achievement and forward program to optimize asset performance.
5. Site acceptance stage gate review.

5.7 CHAPTER SUMMARY

This chapter provides an insight into

- Realistic and achievable project plan development and the use of different formats, including the scheduling of detailed work packets during installation and commissioning
- Managing project teams, including how to facilitate/coach teams and create a great working environment so that they can achieve more
- Managing risk, including the identification of hazards, the timing of risk assessments, and the use of witnessed inspections to manage the glide path to flawless operation
- A project leader's guide to EEM project delivery
- Links with other chapters, including specification management project governance

The key learning points as are follows:

- Use milestone plans supported by detailed activity planning to guide projects through the EEM process.
- Set exit criteria for each milestone stage gate.
- Assess risk and use witnessed inspection to confirm the achievement of the project quality plan.
- Project managers need to become project leaders to get the best out of the core and stage gate teams.
- The glide path to flawless operation should be managed in a similar manner to a control tower bringing an aircraft into land or a space academy landing a rocket ship on the moon.

6

Project Governance

This chapter covers the fourth and fifth EEM subsystems introduced in Chapter 2, "The EEM Road Map," which concern

- The quality assurance of each EEM step to prevent problems from being transferred to the next step
- The collation and codification of knowledge to guide the delivery of low life cycle cost (LCC) designs

There are close links between the content of this chapter and the topics covered in Chapter 7, "Implementing EEM," and Chapter 8, "EEM Early Product Management"—in particular, how to

- Mobilize and support outstanding capital projects as a vehicle to integrate and improve internal processes, develop talent, and accelerate the delivery of strategic goals
- Deliver step-out product and equipment capabilities
- Systematically reduce time to market

6.1 IMPACT OF INTERNAL PROCESSES

Project governance and business governance capabilities are closely linked. Both involve planning and leadership activities, organizational development, execution, and action/review processes (Figure 6.1).

That means that weaknesses in business governance contribute to weaknesses in project governance. Common symptoms of poor business government processes include

- Difficulties in releasing experienced personnel to support the project
- Delays in decision-making
- Poor cross-functional cooperation
- Gaps in communication and/or knowledge management

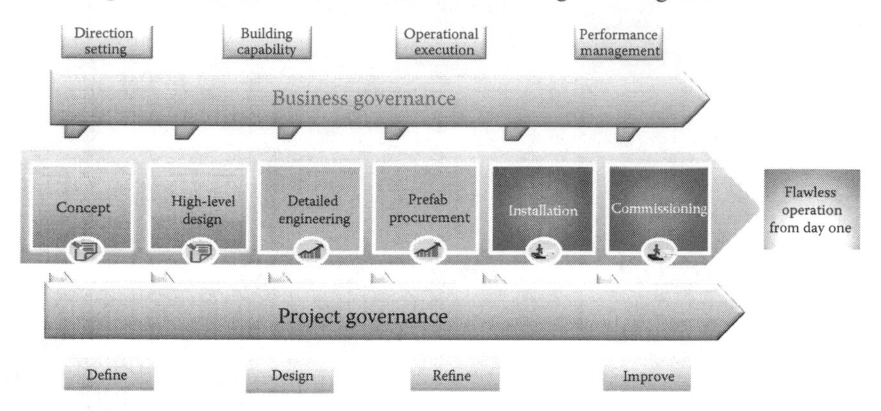

FIGURE 6.1
Business and project governance scope.

These governance weaknesses have contributed to project failures from as early as 1628, when the technologically advanced Swedish flagship *Vasa* sank on its first sailing, killing 50 sailors.* A catalogue of design modifications during building meant that the standard test of stability (30 sailors running from side to side to rock the boat) was cancelled because it showed the vessel to be unstable. The launch of this vanity project went ahead because the king was impatiently waiting for his new superweapon.

More recent projects have suffered also. For example, the Mars Climate Orbiter (1998) was lost in space thanks to incompatibilities with the navigation system because parts of the project used imperial measures and others metric. More recently, in 2015, the French railway company SNCF spent $15 billion on a new fleet of trains that were too large for the stations they were supposed to service. It turns out that the trains were also too tall to fit through some of the tunnels in the French Alps. Ultimately, all of these failures are due to a failure in project governance.

This chapter covers project governance best practice and how business governance can be improved by using project governance as a vehicle to improve internal processes and disciplines. Key areas include

---

* Eric Scigliano, 10 technology disasters, *MIT Technology Review*, June 2002.

- The project governance mind set and skill set (see Section 6.1)
  - How investment decisions are taken
    - Strategy-driven versus reactive management
  - Collaboration with strategic partners
    - Developing the capability to collaborate across company boundaries, accelerating growth from products/service development and access to new markets
- Organizational development (see Sections 6.2 and 6.3)
  - The use of projects as a way to develop talented and high-performing teams
    - Team mobilization, setting accountabilities, supporting delivery, and recognition as part of the EEM stage gate review process
    - The stage gate role as a guide and mentor to the core team
  - Developing and implementing improved ways of working
  - Delegating and empowering front-line teams
- Policy deployment (see Sections 6.5, 6.6, and 6.7)
  - Cascading the insight gained during research into new assets into streamlined ways of working and into problem and defect prevention across the organization

The chapter concludes by setting out the project governance tasks during the *define*, *design*, *refine*, and *improve* phases of the EEM road map introduced in Chapter 2.

6.2 PROJECT GOVERNANCE: MIND SET AND SKILL SET

When an international manufacturer was acquired by a U.S. company, it was given access to funds to reshore part of its product range to the United Kingdom from China. This had been a pet project of the management team for some time, but their previous venture capital owners were not prepared to make funds available to do it. Keen to show its new owners that it had made the right decision, the management team released its best engineers and operations personnel from current operations to work on the capital project.

Within a month its waste levels in the current operations had doubled and service levels had dropped alarmingly. This was almost entirely due to a lack of standard practices and the high complexity of the current operation. The apparently stable process had required almost continuous experienced oversight, and the fact that scrap could be blended with virgin material feeds masked the problem. The company had no choice but to divert project team members back to operations. Unfortunately, the loss of expertise at a crucial time in the project contributed to over six months of start-up problems.

The root causes of these issues can be traced to weaknesses in internal processes and the governance of day-to-day operations. The problems caused by the release of experienced, capable people showed up weaknesses in working method design, skill development, and succession-planning processes, weaknesses that best-in-class organizations learn to overcome.

A survey* of 200 executives from leading UK and international organizations indicates how the internal management processes of industry-leading organizations help them to avoid these pitfalls. The research provides an insight into the differences in management behaviors between those performing less well in quartile 1 and the industry leaders in quartile 4, as shown in Table 6.1. The survey was undertaken

TABLE 6.1

Characteristics of Industry Leading Performers

Strategic Management Processes	Management Outlook	
	Quartile 1 (Low)	**Quartile 4 (Industry Leaders)**
Set direction	Ad hoc problem-solving when required (reactive)	Business-led improvement goals aligned across functions
Build capability	Emphasis on individual training and functional competence	Emphasis on working with partners to fuel growth and delegation to high-performance teams
Performance management	Emphasis on output and cost control	Emphasis on problem prevention and focused improvement
Results delivery	Top-down-directed improvement projects allocated to individuals to complete alongside their day job	Time allocated to cross-functional team-based improvement processes that engage all levels simultaneously

* Hidden Factory survey carried out by DAK Consulting, 2012–2013.

to identify how the management processes of exemplar award-winning organizations differed from those of less successful organizations. Those taking part in the survey attended site visits to 15 exemplar and award-winning UK organizations. Participants compared their organizational characteristics with those of the exemplar organizations using a simple set of definitions.

The survey profile is not presented as representative of all organizations. Due to the nature of the survey conditions, it is likely to contain more organizations proactively seeking improved ways of working. The results are presented here to illustrate how differences in outlook impact on project governance capabilities.

An analysis of the 15 exemplar organizations over the previous five years showed that they achieved around 30% higher levels of effectiveness and profitability than the industry average. In addition, these 15 exemplars had similar journeys, each adopting a systematic small-step improvement approach to working methods and process improvement.

During their journey they leveraged increased stability from performance improvement, delegating routine activities to front-line teams. This released time for experienced personnel to support projects aimed at growth and improved customer value stream performance. The most advanced included cross-company teams working on value chain improvements. The survey concluded that a significant strategic characteristic of quartile 4 organizations is a higher level of cross-functional teamwork and collaboration.

The business governance capabilities of best-in-class industry leaders includes the ability to manage cross-functional teams, building their internal capability to match their strategic challenges. This combines with the ability to apply front-line lessons learned to future strategies to complete the strategic control feedback loop (Figures 6.2 and 6.3).

These capabilities help such organizations to achieve higher scores on the project governance maturity index in Table 6.2.

The project governance maturity for the case study company was assessed at 2 out of 5 using the scale set out in Table 6.2. It is no surprise that in this environment, weaknesses in investment planning, the collation of design guidelines, and risk assessment processes went unnoticed.

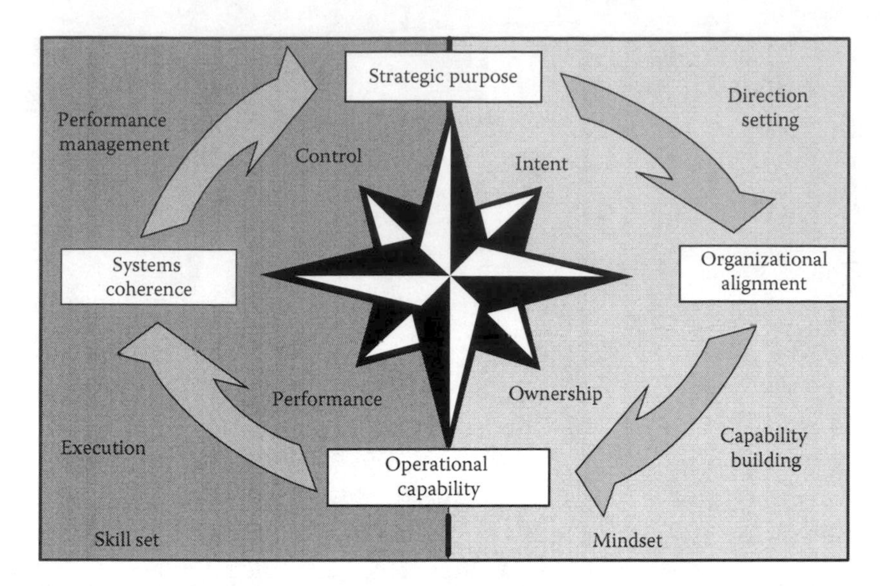

FIGURE 6.2
Strategic compass.

TABLE 6.2
EEM Project Governance Benchmarks

Project Governance Maturity		
1. Weak	**3. Acceptable**	**5. Excellent**
Senior management governance limited to financial oversight and ad hoc involvement to deal with problems as they occur	Formal stage gate review process in place for each EEM step Supported by audit/coaching to develop cross-functional teamwork and project delivery capabilities	EEM stage gate process including annual shutdowns, new products and services, and IT systems Projects seen as a vehicle to improve cross-functional collaboration and raise site capability

To avoid such project problems in the future this company would need to improve internal business governance processes to strengthen the feedback between

- Strategic goals and focused improvement priorities (see the equipment master plan description in Section 6.1.1)
- Equipment performance and investment decisions, so that investment is targeted to those areas that cannot be delivered in any other way
- Investment briefs and project delivery road maps, which reveal risks and opportunities to increase project value

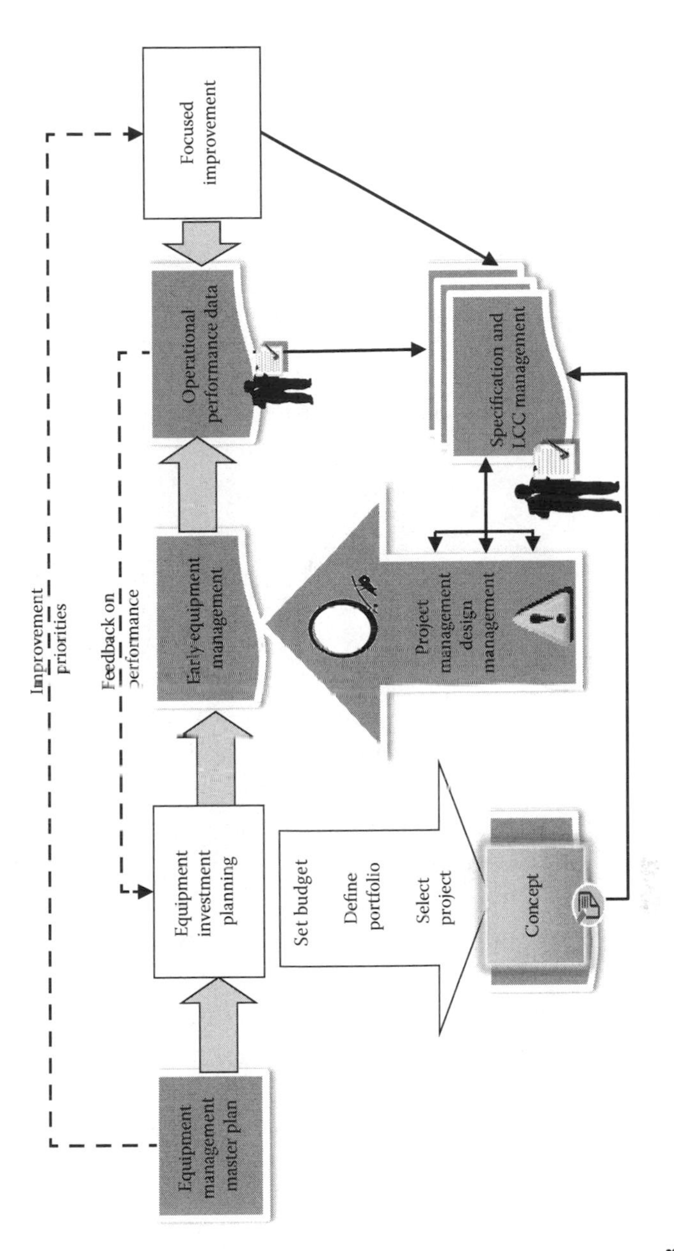

FIGURE 6.3
EEM strategic feedback loops.

Implementing these business process improvements as part of the project delivery process would enhance both project results and enable progress toward best-in-class operational performance. This is discussed in more detail in the following sections.

6.2.1 Direction Setting

Purpose: To develop a forensic understanding of the value of alternative uses of resources/time and guide those resources to achieve their full potential

> At the early stages of an EEM project for a major supplier to the aerospace industry, an EEM team identified that instead of spending almost £10 million on additional presses, actions to improve the condition of the existing plant would provide the additional capacity.
>
> A similar assessment at a soft drinks bottling plant identified the opportunity to avoid capital investment of around £5 million.
>
> In both cases, the original investment decisions were taken as a knee-jerk response to predictable shifts in market forces.

Both of these organizations had sophisticated financial management processes, neither organization had linked these to an *equipment management master plan* setting out the links between technology and business development goals (Figure 6.4).

The process of updating the equipment master plan helps to align the outlook of the three EEM stakeholders: commercial (customer facing), operations, and technology. Integrating and updating their plans as part of the strategic management process reduces the risk that investment decisions have to be made on the fly. The equipment master plan also provides the opportunity for front-line and middle managers to communicate upward so that senior management are fully informed on the progress of current equipment performance, improvement goals, and the consequences of potential future investment decisions.

The interface between investment planning, EEM project governance, and updating the equipment master plan is set out in Figure 6.5.

6.2.2 Capability Building

Purpose: To develop organizational skills so that equipment and processes are fully understood, under control, and flexible to shifts in demand

The process of updating the equipment management master plan highlights knowledge and competency gaps and, as part of that, where resources

	How is this changing	Challenges	Equipment management goal
Commercial • Customer • Expectations • New products and services	Increasing expectations of quality cost and delivery performance	Incremental product development, new products and services for growth Supply chain for global operations	Capacity for future demand Robust supply chain Simple logistics/forecasting needs Flexible to potential market shifts
• LCC Management • Main contributors (4M) • Technology	Material costs, energy costs, labor costs, batch sizes	Sustaining volume on existing products while introducing new ones Energy inflation	High level of resource recycling Flexible to financial risks (e.g., Vendor) Easily scalable to 400% or to 25% Access to high added-value markets
Operations • Operability • OEE losses • Lead time/flow/flexibility	Focused improvement smaller batches, new product life cycles, shorter lead times	Systematic hidden loss and lead time reduction Achieve "normal operations" with less effort	One touch operation for height, position, number, color, etc. Flexible to volume risk Flexible to labor skill levels
• Maintainability • Predictable component life • Defect prevention	Reduced plant availability for maintenance, more complexity Increased precision, availability of skills	Extending MTBF/MTTR Maintenance prevention Supporting process optimization Lack of skills	In-built problem diagnostic Self correcting/auto adjust Predictable component life Easily overhauled, fit and forget
Technology • Reliability • Increase precision • Automation • New technology/materials	Improved control, lower grade Materials	Capturing lessons learned Adapting plant to new products, improving technical competence	High MTBI Stable machine cycle time Easy to measure Flexible to material variability
• Safety • Ease of compliance • New legislation	Increased legislation and controls Higher environmental expectations	Safe, environmental practices with minimum impact on performance Improved sustainability	Foolproof/failsafe operation High level of resource recycling Uses sustainable resources

FIGURE 6.4
Equipment management master plan challenges.

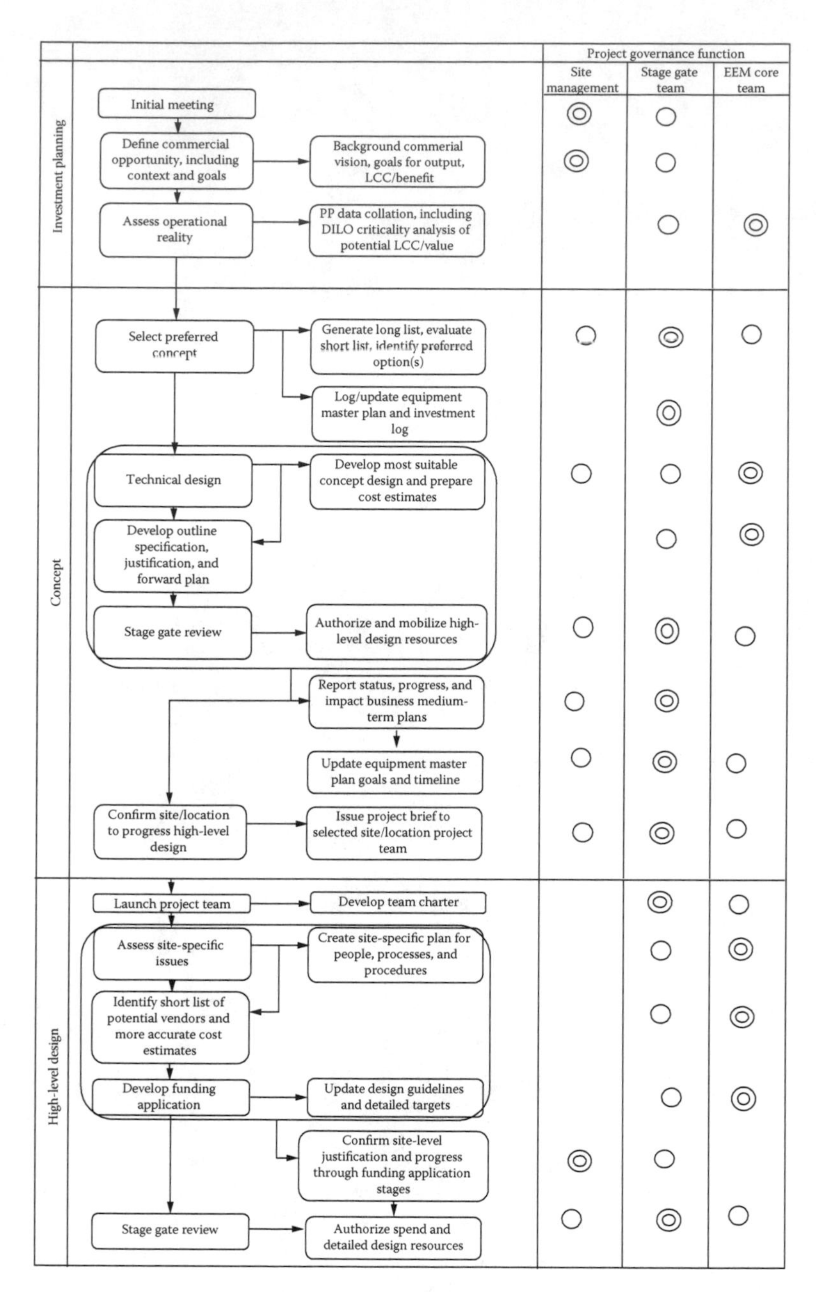

FIGURE 6.5
Investment plan and equipment master plan update.

need to be targeted. This sets out where specialists and key managers need to engage with higher value-adding challenges. That provides an incentive for those at the top of their game to delegate more and an opportunity for personal development both for the specialist and the person backfilling their role.

People learn faster through practical application, particularly when it is supported by on-the-job support and coaching. Coaching and support for project team members is an important part of the project governance process.

With new equipment comes the opportunity to bring in new technology that can create gaps in internal skills. For example, many vendors are replacing mechanical linkages with electronic controls. Maintenance engineers that have developed their experience with predominantly mechanical devices may need to develop new competencies. In addition, the experience of the aerospace industry has been that increased levels of automation and self-correcting devices have reduced the level of pilot awareness of the onset of potential failures. That means that the engineer has less information about how well the aircraft has functioned. In this environment, condition monitoring and process trending become more important. To many, these are gray areas at the extremes of their knowledge. Without the focus on building capability, these gray areas can go unnoticed, storing up problems for the future.

6.2.3 Performance Management

Purpose: To develop the insight to control execution processes, coordinate resources, and prevent future problems

Vendor Performance

A key area of project performance management is linked to vendor relationships. Traditionally, this has been driven by cost and ultimately attempts to minimize or protect supplier margins. The EEM focus is on reducing project LCCs, driven by maximizing project owner margins.

This is in contrast to traditional fixed-fee lump-sum contracts, where

- Purchasers attempt to pass on risk/unknowns to vendors backed up by stiff penalties for nonachievement.
- Vendors submit lower costs knowing that there will always be opportunities to improve their profitability due to unforeseen (but usually very predictable) circumstances.

When problems occur, one of two things happens.

1. Vendors finding that profits are going to be hit take steps to minimize their costs by cutting corners.
2. Purchasers divert resources into taking potentially protracted legal action.

The outcome is project delays and everyone, except the legal profession, suffering losses.

In reality, vendors are only able to directly impact a portion of the factors that contribute to project success/failure because the achievement of flawless operation means zero sporadic losses or no breakdowns on day one. This is achieved by managing the impact of 1) equipment conditions and 2) operational practices and human error.

It may be possible for the vendor to directly influence equipment design and delivery, but they have only indirect influence on the development of operational competence to prevent human error. In addition, if there is complexity in current product recipes/quality assurance (QA) methods, weaknesses in supply chain controls, poor business processes, or low skill levels, the impact of these will be magnified during the early stages of the new operation.

Central to EEM is a proactive approach to risk management where stage gates are used to capture problems early and make visible opportunities for additional project value. An example of how this can be incorporated into commercial terms is in the approach used by BAA as part of the Terminal 5 project. Although this project hit the headlines due to problems during start-up, the design and construction of this £4.2 billion megaproject are generally considered a success.

Experience had taught BAA that actively managing the causes of risk was more successful than trying to protect themselves against the effect of risk. To achieve that success, BAA replaced the negative content of traditional, often confrontational contracts (who pays when things go wrong) with a commercial model agreement and commercial policy that creates commercial tension but not a commercial barrier (Figure 6.6).

BAA agreed the direct project costs with the supplier and the profit level they were allowed to make on top of that. BAA held all the risk and actively managed these risks. The vendors profit margin was guaranteed, so they would not lose out if things went wrong. In return, contractors

- Open-book accounting
- Contractors profits ring-fenced
  - For an agreed scope of work and agreed level of resources
  - Payment based on time and materials
  - Client has the right to audit actual costs at any time
- Incentive plan with potential to increase contractors profits
  - Agreed target costs and client aspirations (acceptable/excellent)
  - Benefits of exceptional performance shared 50/50
- Compensation for change
  - Nil if within design scope
    - Evolution of plans and adaptation to circumstances is an inherent part of large projects
    - Actual costs reimbursed but profit not increased
  - Increased compensation for change of scope
    - Revised resource plan and ring-fenced profits

FIGURE 6.6
T5 vendor contract terms.

were expected to work in partnership to assure success and manage failure risk. Contractors were also expected to provide a high level of transparency so that BAA could audit their books to confirm what was actually spent.

Contractor teams shared project liabilities. In the event that an omission was found that was due to an oversight on the part of the vendor, the vendor dealt with the problem at cost. That is, their profit margin was protected but they did not earn more profit on the additional work. If additional work was identified that was not anticipated when contracts were placed, this would be charged at cost plus an agreed margin.

BAA created an incentive fund from targets on all projects. BAA tries to price risks and opportunities and agree a price with suppliers to incentivize active risk/opportunity management without introducing pain. The incentive fund is there to encourage suppliers to better the agreed plan. It is a means for suppliers to increase their profit provided they perform and respond to nonperformance risks.

Anecdotal evidence shows that companies that succeed with projects have double the amount of informal contact at peer-to-peer levels than those that fail. During that increased contact, internal resources learn more about how to get the best out of the new assets. They also increase their vendors' understanding of what is needed so that latent design weaknesses are identified early on in the project and dealt with when they cost less. From the vendor perspective, they receive new product development input direct from the customer and a great reference site to support future business opportunities.

It is difficult to point to evidence of the impact of collaboration with vendors during projects. Increased collaboration in manufacturing supply chain relationships has been shown to achieve up to 10% advantage in service levels with similar or even lower logistics costs. Most benefits come from reduced material costs, improved customer service, and faster growth in market share. Other benefits include improved greater flexibility to change and faster translation of opportunities into revenue. There is no reason to doubt that similar additional benefits are achieved from improved collaboration with project partners.

Collaboration with vendors is a fundamental part of the EEM process. This does not mean that contractual terms are any less rigorous than in the traditional approach. The foundation of an effective working relationship is that each party is clear about what is to be delivered and at what cost.

6.2.4 Results Delivery

Purpose: Work routines that convert plans into reality

The project governance role concerns the assurance that each project step will achieve flawless operation from day one and deliver the full investment potential. Figure 6.7 sets out the results delivery processes to support that role.

This illustrates how project governance results are dependent on the effective coordination of

- Site equipment master plans
    - Performance improvement targets
    - Awareness of relevant technology developments
    - Feedback on current asset design weaknesses as design guidelines
- Best-practice design books
    - The capture and codification of tacit knowledge
    - Up-to-date technical library and procedures
- Design and performance management
    - Stage gate–based product and project delivery processes capable of delivering flawless operation from day one, including
        - Innovative, friendly diverge/converge option evaluation
        - Preventing the transfer of problems to the next stage

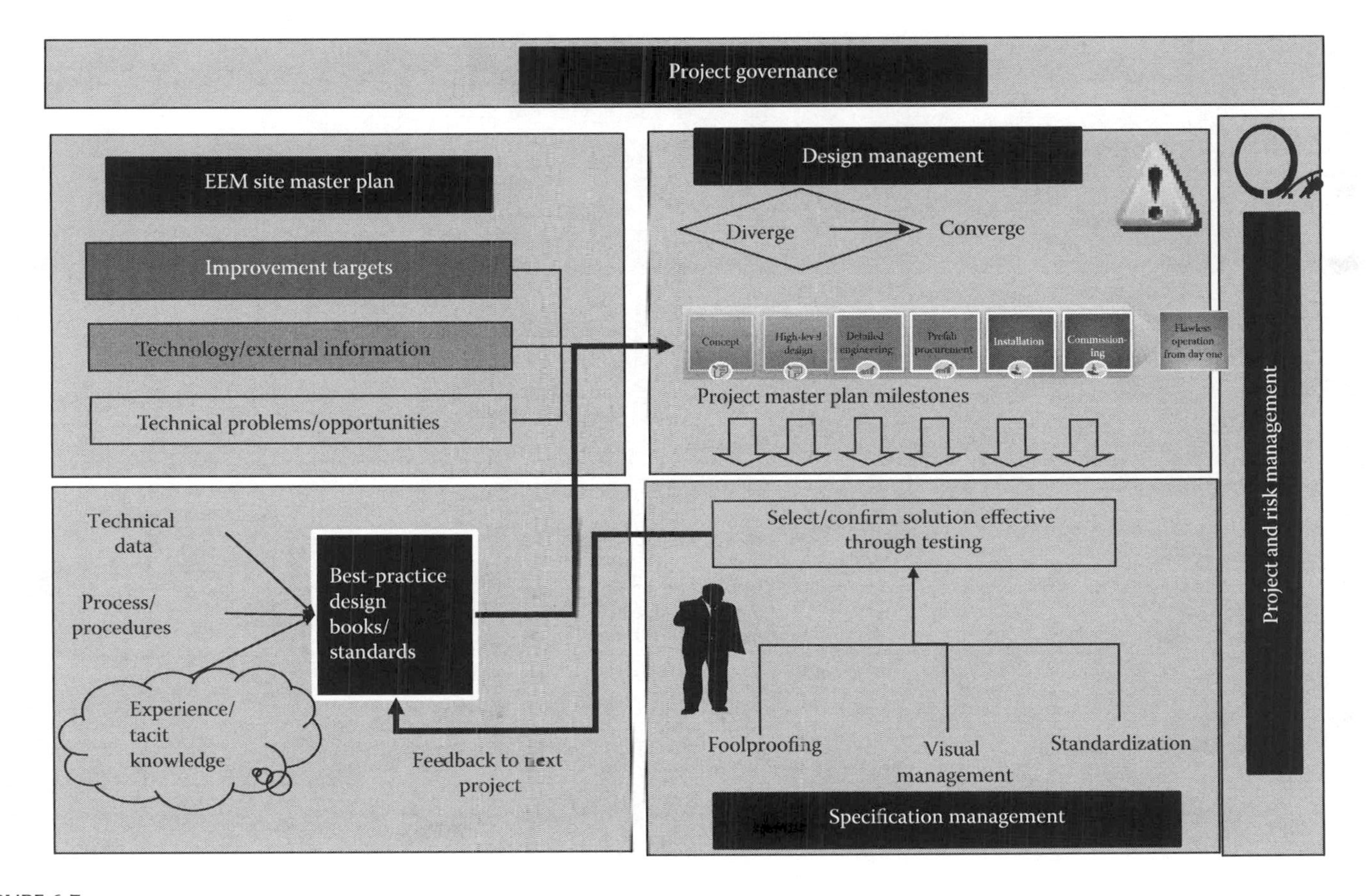

FIGURE 6.7
EEM result delivery processes.

- Specification and LCC management
    - The use of objective testing processes to refine preferred options using design goals, guidelines and standards, visual management, and foolproofing
    - LCC model application
- Project and risk management

These four subsystems are like links in a chain supporting the project governance tasks at each step of the EEM road map (Figure 6.8). The achievement of flawless operation is a significant challenge (Figure 6.9 and 6.10) but these gains can be lost over time unless project governance achieves a seamless hand over to business governance.

To lock in the gains and deliver the full potential of the investment day one operation should be viewed as a launchpad to optimized operations. Optimization is part of the *total productive maintenance* (TPM) journey toward zero defects. The journey is characterized by actions to identify and systematically reduce minor product defects. These are defects the customer would not notice but which indicate that the process precision is not 100%. Understanding how to control process variation leads to less adjustment at changeover and less intervention during steady state and close-down. It is worth noting that extending the mean time between interventions from minutes to hours breaks the link between direct labor hours and output. The role of direct labor is profoundly changed from that of loading and unloading to that of coordinating the value stream. The increase in precision and flexibility opens up the potential for smaller batches, increased variety, and the use of cheaper materials—all important areas of competitive advantage.

This process of defining and delivering optimized operations cannot start until a process is stabilized; hence, this is a feature of flawless operation. Furthermore, if the glide path is successful, during the process of testing and confirmation, project resources will have the time to focus on defining the optimization road map so that this is part of the package handed over at site acceptance testing.

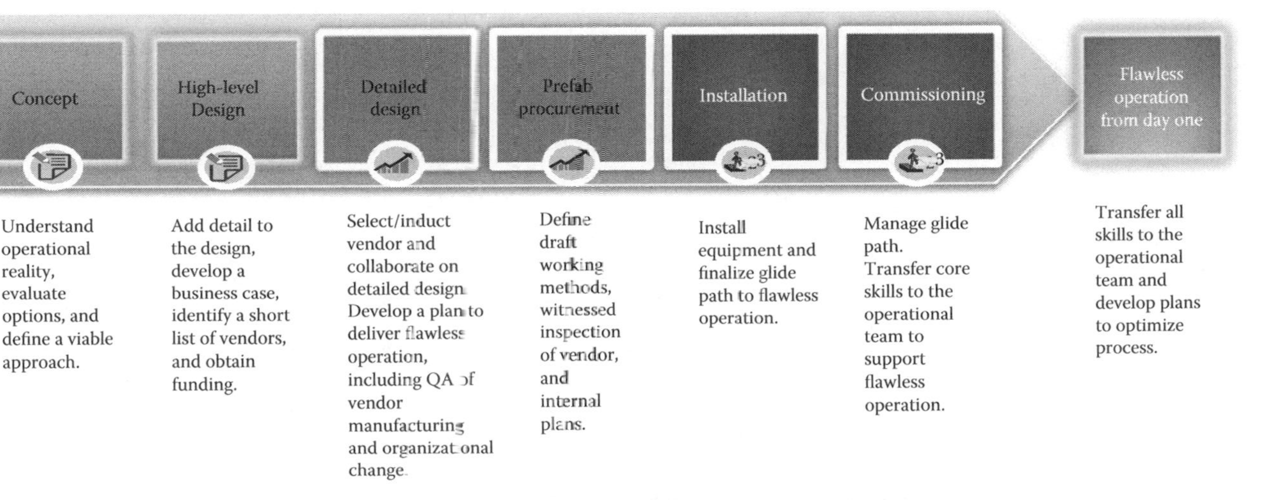

FIGURE 6.8
EEM project governance oversight.

$$\text{Vertical start-up \%} = \frac{\text{Actual start-up time}}{\text{Planned start-up}} \times 100\%$$

Project Scope	Flawless Operation
Single simple machine	< 1 day
Single critical or complex process	< 1 week
Full line	< 1 month
New site	< 1 year

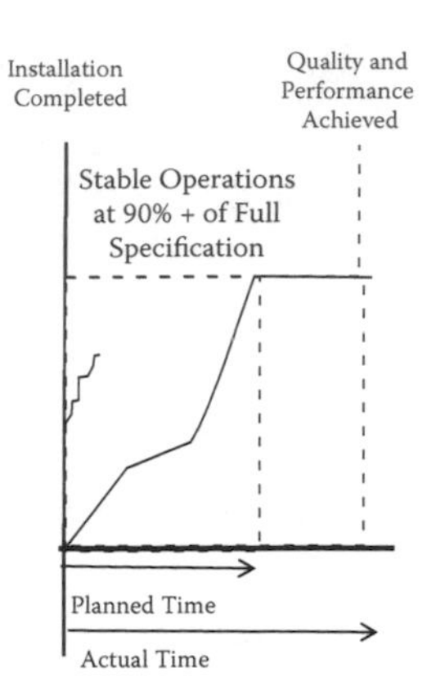

FIGURE 6.9
Flawless operation benchmarks.

	Unit	Definition
Flawless operation	%	Actual vs. planned
Audit score	%	Actual audit score
LCCR life cycle cost Reduction	£/year	Comparison of base case, standard design, and installation
No of EEM improvement	Qty	Improvements identified and implemented during project as a result of EEM
Conformance to budget	£	Actual expenditure vs budget
Project OTIF	%	Achievement of milestone quality criteria on time in full
Witnessed inspection results	%	Test completion right first time
Customer satisfaction project management	%	Based on response to questionnaire

FIGURE 6.10
EEM project key performance indicators (KPIs).

6.3 ORGANIZATIONAL DEVELOPMENT: LEADERSHIP CHALLENGE

EEM is a continuous improvement process for project delivery. Each project builds on the lessons learned from previous projects.

Although EEM was developed for capital projects, it is equally relevant as a delivery vehicle for any project including software application,

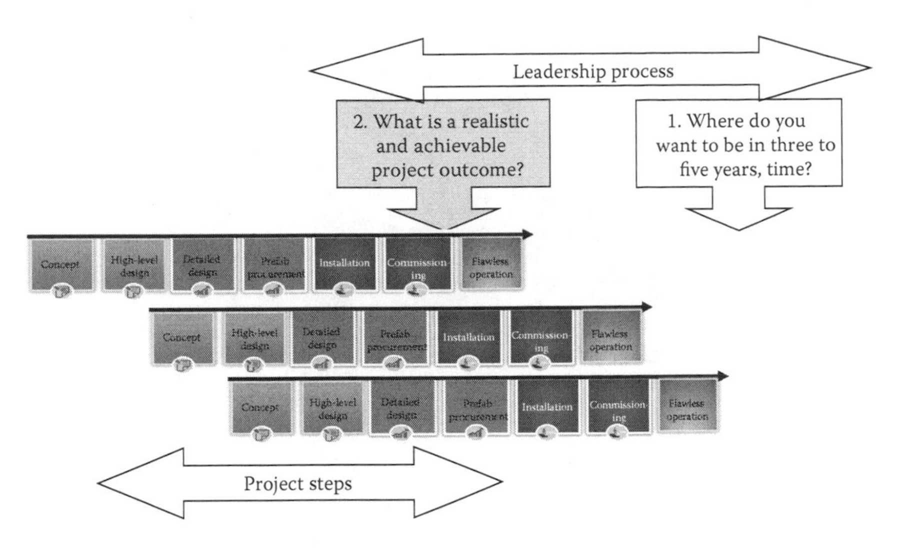

FIGURE 6.11
Setting the leadership agenda.

shutdowns/overhauls, new product development, business start-ups, and so on. Project delivery inevitably introduces change and an essential part of successful change is leadership. As all organizations undertake multiple projects each year, the governance of these is undertaken within the context set by the internal leadership agenda (Figure 6.11).

The governance challenge is to recognize and deal with barriers to project success before they impact on project results. Alternatively, identify the root causes of failures and prevent reoccurrence. One of the advantages of leadership activities with project teams is that each team member is working outside of their normal environment. Each member of the cross-functional team will need to establish new working relationships.

A formal team induction process presents an ideal opportunity to establish a proactive outlook and challenge limiting behaviors. Reinforce this with coaching support during each project step, and as the project team touch more of the organization, they will influence the values and outlook of those who take over the new operation (Figure 6.12).

Further discussion about the mechanics of developing the cultural outlook is included in Chapter 5, "Project and Risk Management."

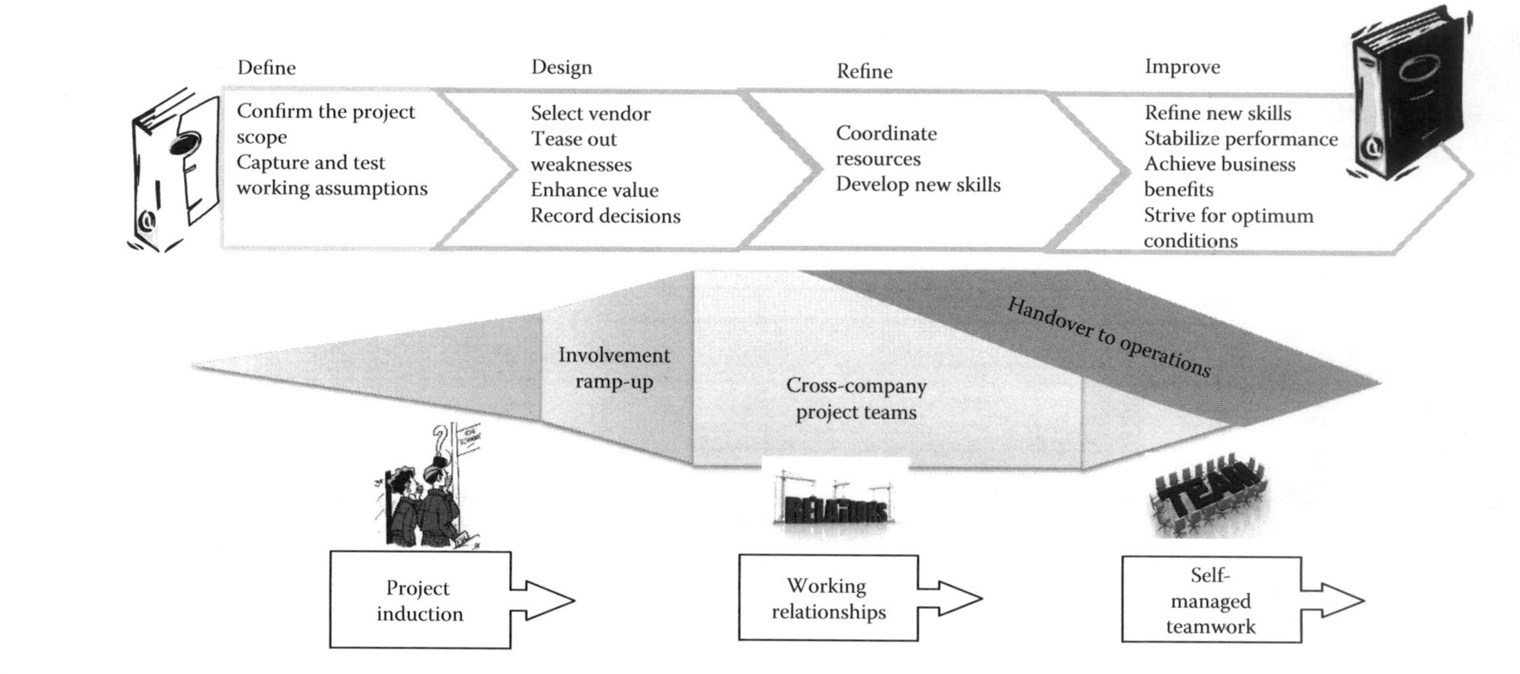

FIGURE 6.12
Shaping project and operational culture.

6.3.1 Setting EEM Policies

One of the most important leadership roles is to clearly define accountabilities, to be able to:

- Hold people accountable.
- Provide recognition of successful outcomes.
- Reinforce good practices and challenge limiting behaviors.

6.3.1.1 EEM Policy Example

Chapter 2 set out RACI charts for each role in Figure 6.13.

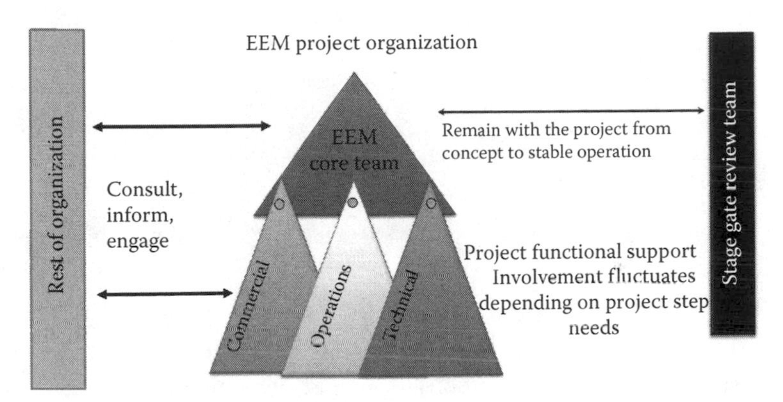

FIGURE 6.13
EEM project roles.

The precise definition of EEM roles will vary from company to company. The following EEM policy example was developed by a UK-based subsidiary of a multisite international food and drinks supplier.

6.3.1.2 Policy and Direction

- At a corporate level, the *EEM steering committee* will be responsible for setting and deploying EEM policy.
- The site equipment master plan will be maintained as part of the site manufacturing plan.
- At project level, a *stage gate review team* will be selected from the site management team to take responsibility for EEM project governance to assure the rigor of the stage gate reviews, including planning, resource allocation, and the removal of roadblocks to progress.

- Within the site-level improvement program, EEM will be incorporated into the *focused improvement pillar*.
- The control of EEM process methodologies, templates, and underpinning systems will be managed by nominated subject matter experts.

6.3.1.3 Project Delivery

The EEM core team will be accountable for applying EEM principles to the project and adapting/refining underpinning systems as necessary to meet EEM project goals. The EEM core team members will become leaders of functional representatives/teams formed to support the core team activities.

6.3.1.3.1 Commercial Core Team Member Role

Accountabilities include

- The identification of design guidelines for LCC drivers and customer value features
- Logistics coordination
- Communication/progress coordination
- Leadership of the concept and basic engineering steps

6.3.1.3.2 Operations Core Team Member Role

Accountabilities include

- The identification of design guidelines for
  - Operability (ease of use)
  - Maintainability (ease of maintenance)
- Leadership of the commissioning stage onward

6.3.1.3.3 Technology Core Team Member Role

Accountabilities include

- The identification of design guidelines for
  - Intrinsic reliability issues
  - Safety issues
- Updating the EEM knowledge base
- Leadership of the detailed design and construction stages

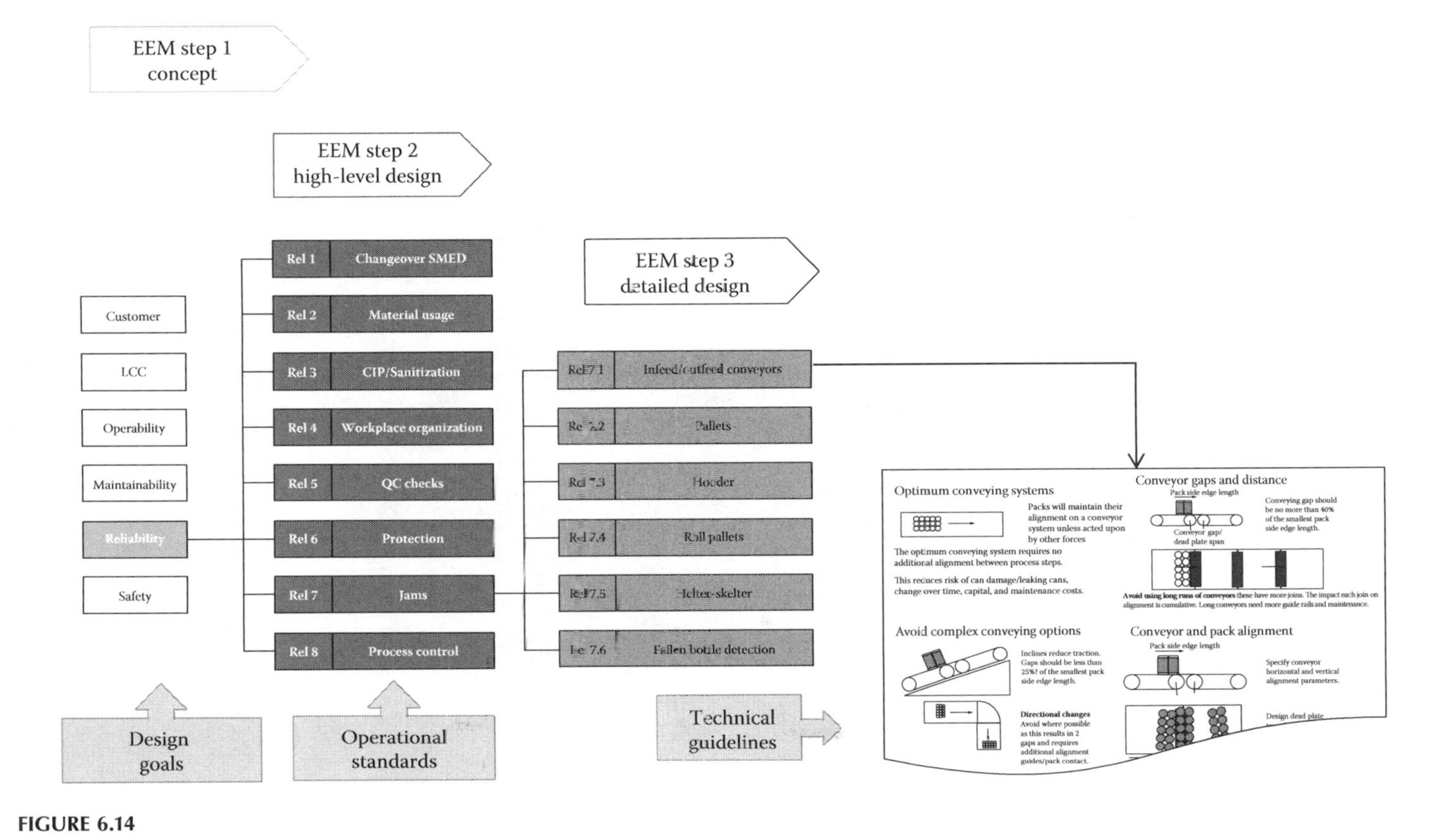

FIGURE 6.14
EEM design goal and standard hierarchy.

6.3.1.4 EEM Support

6.3.1.4.1 EEM Subject Matter Expert

- Responsible for supporting the project team's application of EEM principles and techniques

6.3.1.4.2 Project Manager/Installation Manager

- Responsible for ensuring that EEM policy/methodologies are applied
- Works with the focused improvement pillar champion to apply EEM principles to project proposals

6.3.1.5 Best-Practice Design Book

Ownership will be distributed across site engineering personnel in line with current asset management protocols. Their role is to update the design goals, guidelines, and standards with lessons learned from this and other projects (Figure 6.14).

6.4 ORGANIZATIONAL DEVELOPMENT: MEASURING PROGRESS

6.4.1 Setting Targets

The starting point for setting design targets is an analysis of current performance and improvement potential. The scope of this analysis should include not only the asset under review but also the impact of upstream processes and on downstream processes.

Figure 6.15 contains an assessment of overall equipment effectiveness (OEE) over a representative period. This includes a design target based on *best-of-the-best* OEE performance over the period. This is calculated using the best availability, performance, and quality results over a representative time period as an indication of how a new asset would perform. Some may consider that new assets should be 100% effective. The reality is that equipment performance is impacted by factors such as material consistency, demand fluctuations, and planning constraints. These will not disappear simply because a new asset has been procured. In fact, if the new asset has a higher capacity, it may result in additional planning constraints to upstream or downstream processes. The best-of-the-best OEE assessment

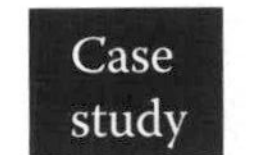

OEE analysis over representative period

Automotive

	Availability	Performance	Quality	OEE
Jan	93.70%	63.70%	99.50%	59.4%
Feb	91.70%	64.90%	99.50%	59.2%
March	90.40%	63.80%	99.60%	57.4%
Average	91.93%	64.13%	99.53%	58.7%
BoB	93.70%	64.90%	99.60%	60.6%
Gain	1.77%	0.77%	0.07%	1.88%
Target	91.93%	85%	99.53%	77.8%

From average to best of the best
OEE plus improved downtime = 10.03%

Need to identify additional 6.9%
improvement using value
engineering to hit target of +17.2%

	Asset	Ref period downtime	Criticality
1	AID	2.1%	18
2	Drier	1.6%	21
3	Laser	1.0%	18
4	Infeed conveyor	.75%	16
5	X-ray	.30%	17
6	Pick and place (focus on robot issues)	2.4%	Robots 14
7	Ovens		15
8	Wash coat to line		15
9	Networks		13
10	Bar code readers		12
	Total	8.15%	

FIGURE 6.15
Mapping performance target gains.

is a reasonable basis for setting realistic and achievable OEE targets for new equipment. In this example, the project has been justified against an effectiveness target of 77.8%, compared with a current average of 58.7%. With a best-of-the-best OEE analysis indicating new assets would achieve 60.6% under current operating conditions, the analysis made it clear that improvements to factors external to the new asset were needed. As this analysis shows, following the best-of-the-best assessment, the team went on to identify other areas of hidden loss that then became part of the project scope.

6.4.2 Beyond the Stage Gate Review

The EEM road map makes use of milestone planning to define the main signposts on the journey. Progress at these signposts can measured against milestone QA completion criteria. The EEM road map sets out six milestones that apply to all projects.

Figure 6.16 shows the format for a project quality plan showing the first three steps of the EEM milestone plan.

Work can start on any milestone at any time but a milestone cannot be completed until the completion criteria for all previous milestones have been achieved. That means that an EEM audit process can be used to confirm that the completion criteria for each EEM step has been completed as part of a stage gate review at the end of each step (see Section 6.3.4).

The stage gate review process is the most visible part of the project governance role. Through this event, at the end of each step the stage gate team can

- Manage the evolution of design and specification development processes
- Prevent risks from being transferred to later steps and support the project management processes

Feedback at the stage gate review is carried out by the EEM core team, who explain

- The progress made since the last review, including an assessment of potential project problems by each of the commercial, operations, and technology teams, covering
  - Red flag status: Issues that have a significant impact on future project success. The stage gate cannot be passed without resolving this issue.

EEM project

No.	Results paths: Procedure	Process	People	Program overview: Milestone
1				Start
2				1.1 Mobilize concept
3				1.2 Select preferred concept
4				1.3 Create concept specification
5				1.4 Develop project plan
6				1.5 Concept stage gate review
7				2.1 Mobilize HLD
8				2.2 Select concept delivery approach
9				2.3 Define HLD specification
10				2.4 Develop project plan step 2
11				2.5 HLD stage gate review
12				3.1 Procure
13				3.2 Detailed design
14				3.3 Freeze specification step 3
15				3.4 Detailed activity planning step 3
16				3.5 Detailed design stage gate review

FIGURE 6.16
EEM milestone plan overview.

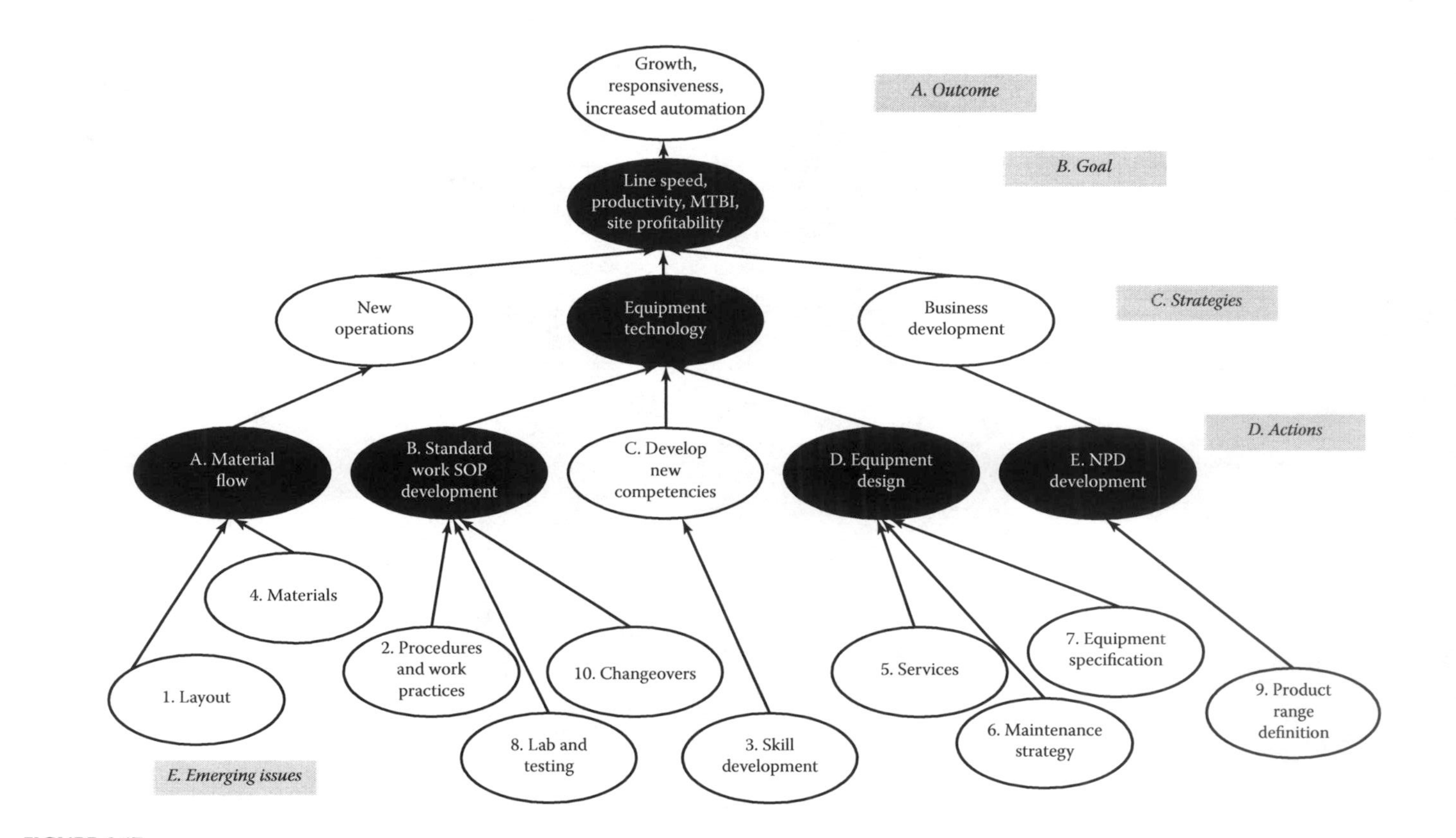

FIGURE 6.17
Action map example.

    - Amber flag status: Issues with a known route to resolution that are unlikely to have significant impact on the delivery of project goals or working assumptions based on experience but in need of confirmation. These items must be completed prior to the next stage gate review.
    - Green flag status: Critical issues that have been resolved satisfactorily.
- Action plans and resources needed to complete tasks up to the next stage gate.

The timing of the review is one to two hours, with roughly half of this time spent on progress to date and half on forward plans.

Any of the commercial, operations, or technology teams can prevent a project from passing a stage gate by raising a red flag issue. A red flag issue should only ever reach a stage gate review if a suitable compromise cannot be reached. The issue can then be referred to a higher authority as necessary.

6.4.2.1 Stage Gate Team Coaching Role

Although the most visible stage gate team activity is the stage gate review, the project governance role involves more than the six stage gate review meetings. Core team members will need support from the stage gate team and their functional managers to be able to set goals and develop realistic and achievable plans to deliver those goals.

The nature of projects is such that the number of personnel involved fluctuates over the project lifetime, as do roles and responsibilities. Those who are new to projects will need support from their functional managers to be able to operate effectively in this relatively unstructured capital project environment.

Avoid the mistake of releasing personnel to a project team and assuming that their management is then outside of the scope of operational reporting lines. Cross-functional core team members work best when they provide a communications conduit between the project and the rest of the organization.

During the normal day-to-day routine, the direction of information flow is downward. Those involved in day-to-day shop floor reality are the ones who can make or break investment performance. That means opening up channels of upward communication. This includes allowing time and space for managers to reflect and directly report on challenges faced, options selected, and plans made

Project "war rooms" are a great place to do this (see Chapter 5). Here, the key information is visible and the links between issues can me more easily considered. Toyota consider Oobeya (Japanese for "big room"/"war room") to be the backbone of the Toyota Management System. This is where the project sponsors and project teams meet to develop briefs, carry out design reviews, review progress, and agree actions. Just like in a CSI investigation, it is a place to look for connections and solve puzzles.

The project governance benefits of project war rooms include

- Better visibility and engagement
  - Improved communication, visibility of plans, and workload
  - Improved engagement of leaders and teams
  - Ease of priority setting
- Greater accountability
  - Focused targets driving team activity and proactive behavior
  - Clarity of metrics and the alignment of commercial, operations, and technology team goals
  - Increased team awareness of parallel workstreams
- Better cross-functional collaboration
- Quicker problem-solving
  - Everything to hand
  - Problems made visible and shared
  - Creates the conditions for innovation and new thinking

6.4.3 Coaching Parallel Workstreams

The action mapping process set out in Chapter 3, "Design and Performance Management," provides an opportunity to raise project team member understanding of how the commercial, operations, and technology perspectives interact. Coaching plans can include discussions to

- Confirm understanding of how the project supports business strategic plans
- Assure that the workstreams are sufficiently well defined to provide a balanced, collective plan that takes into account the views and aspirations of multiple stakeholders

When coaching with the project action map (see Figure 6.17 as an example), adopt a two-pass approach.

- Pass 1: Team members should be able to explain, starting at the bottom of the map, how the issues will be addressed by the actions and how these combine to support strategies that will deliver the goals and outcomes.
- Pass 2: Team members should also be able to explain, starting at the top of the map, how the outcomes and goals will be delivered by the strategies, how they will be supported by the actions, and how the actions will deal with the issues.

Try not to change the shape of the map. The goal here should be to confirm understanding and, where necessary, refine rather than replace it. Look to add detail to the action briefing notes and ensure that cross-functional accountabilities are understood. Concentrate on the shaded action nodes. These are the ones with more than one arrow in and out and are more complex.

6.4.4 EEM Audit Process

The formal project governance process is the stage gate review at the end of each EEM step. This is used to

- Trap weaknesses and prevent them from passing to the next step
- Confirm working assumptions
- Agree readiness to proceed and plans to complete the next step

As mentioned above, stage gate reviews are not design reviews. They are relatively short, typically a one to two hour process. The first half of the review is to confirm progress, the second half to confirm plans for next steps. The stage gate review itself is important, but equally important is the preparation effort that the EEM core team put in. As part of this process, they reflect on progress, communicate ideas, and develop plans for next steps. They should also confirm the buy-ins of all stakeholders to their forward plans before the review. During these conversations, any missed items or issues will surface. The conversations will also aid communication and increase project engagement and momentum for change.

As part of this preparation, an EEM step audit provides the project governance framework to confirm readiness to proceed by confirming

- The completion of EEM step tasks
- The achievement of anticipated quality plan outcomes
- The progress made to resolve red and amber flag issues raised at the last review
- The buy-in of all stakeholders to progress to next step

Specific audit points for each milestone are included in Sections 6.6, 6.7, and 6.8.

6.4.4.1 Approach

The EEM process audit provides project quality assurance and support for the development of site/team EEM capabilities relating to

1. Project and risk management: A human oriented approach to equipment projects to trap latent problems and risks early using the expertise of the complete company team
2. Specification and LCC management: The systematic collation of equipment management lessons learned and the development of maintenance prevention (MP) data to guide decisions at each EEM step
3. Performance and design management: The systematic identification and analysis of areas of opportunity/hidden losses to reduce project risks and enhance project value

The benchmarks used in each audit are linked to the EEM core team route goals, as set out in Table 6.3.

Project leaders and EEM core teams provide evidence of the achievement of RACI competencies for the current EEM step.

TABLE 6.3

EEM Step Goals

	EEM Step	Goal
1	Concept	To define the project scope and develop a preferred option and forward program to deliver defined business goals
2	High-level design	To clarify the delivery approach, confirm the business case, and obtain funding
3	Detailed design	To select the right partner, add detail to the basic design, tease out latent design weaknesses, prevent problems/risks, and enhance project value
4	Prefab procurement	Readiness to begin installation
5	Installation	Readiness of the glide path to flawless operation
6	Commissioning	The achievement of flawless operation and the development of the route map to optimum conditions

6.4.4.2 *Audit Roles and Timetable*

Audit roles are set out in Table 6.4 together with an outline audit timetable at Table 6.5.

1. A summary of the current status and progress to date
2. Confirmation of EEM goals incorporated in the design and specification
3. An agreed assessment of strengths and weaknesses
4. A forward program including EEM capability development for current/next steps

TABLE 6.4
Audit Roles

Topic	Notes
EEM auditor	EEM specialists or approved auditors
Project leader	Project leader or deputy
EEM core team	Representatives from commercial, operational, and technical teams
EEM policy champion	Management representative and additional stakeholders as appropriate

TABLE 6.5
Audit Timetable Example

From	To	Topic	Who	Notes
0900	0920	Project background	Project leader and EEM pillar champion	Current status, steps taken to date re: project and risk management
0920	1020	Commercial, operations, and technology assessment	EEM core team	Confirmation of completion of current EEM step activity 1. Specification and LCC management 2. Design and performance management
1020	1040	Agreement of audit conclusions	All	Review progress against EEM step benchmarks
1040	1100	Agreement of next steps	All	To address audit gaps or prepare for the next EEM step

6.5 LOCKING IN EEM GAINS

6.5.1 Origins of EEM

EEM was developed as part of the TPM toolbox, the goal of TPM being the continuous improvement of equipment effectiveness. The key word here is *effectiveness*, which measures whether equipment is

- Available when needed (zero breakdowns, rapid setup)
- Capable of full nameplate capacity when running
- Able to achieve right-first-time quality for each and every unit of output (zero defects)

The development of the TPM toolbox took place over more than three decades, during which time it was recognized that capital investment on its own can paradoxically reduce performance, particularly where the project process has not taken on board the lessons of the past.

To avoid this pitfall, the TPM improvement road map incorporates processes to

- Systemize improvement in the effectiveness of the current generation of assets
- Target investment at those areas where performance cannot be further improved with existing assets

The sequence of improvement tactics is set out in Figure 6.18. The first three levels of focused improvement are designed to achieve zero breakdowns/stable operation. The last two reduce the levels of intervention to optimize process capability. This is the domain of EEM.

Together, these five focused improvement tactics raise levels of technical stability and operational resilience. Figure 6.19 provides a summary of

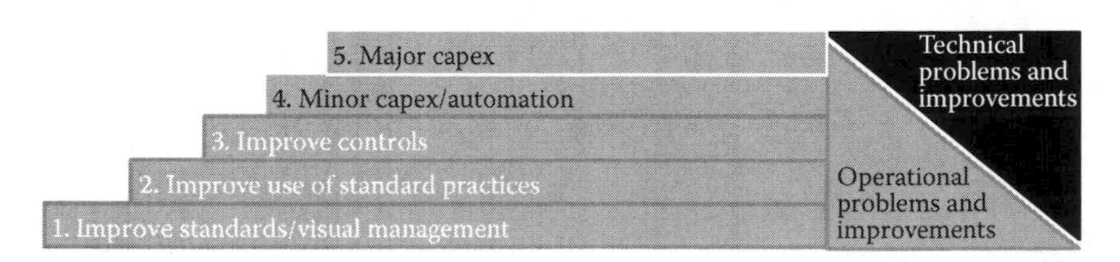

FIGURE 6.18
Focused improvement tactics.

	Technical stability			Resilience		
Score	Operating best practices	Routine servicing	Prevent deterioration	Measure deterioration	Contingency planning	Spares management
1	1. Little/none	1. Limited formal maintenance practices	1. Informal start-up checks	1. Normal conditions not defined, signs of failure generally not acted upon	1. No spare capacity	1. No formal spares policy
2.	2. Documented methods adhered to by most personnel	2. Standard fix to solve procedures used by all	2. Standard asset care practices in place for critical assets. Sources of accelerated wear identified and targeted	2. Normal conditions understood by technical specialists using formal inspection to detect failure conditions	2. Contingency plans in place for all single points of failure	2. Generic control of spares based on overall usage, aim to hold long lead time spares.
3	3. Standard practices in place across all shifts	3. Servicing quality measured and improved	3. Routine audit of compliance results above 90% for all assets, stable component life, accelerated wear/contaminaton controlled	3. Normal conditions able to detect failure conditions as part of routine checks	3. Simple manual activity to restore productive capacity	3. Spare parts plan for every asset
4	4. Asset care integrated into standard practices	4. Systematic improvement in servicing quality and extended time between intervention	4. Systematic achievement of optimum conditions to increase in MTBF	4. Condition based maintenance is succesful at detecting and preventing failures using tools/test equipment	4. Automatic switch over with loss of output	4. Spares plan for every asset linked to CBM
5	5. Working practices are easy to do right, difficult to do wrong, simple to learn	5. Rapid replacement of components possible before failure	5. Zero breakdowns achieved, significant reduction in defects	5. Condition based maintenance using look feel and listen inspection tests against standard possible "at a glance" by all	5. Automatic switch over no loss of output	5. Spares plan for every asset linked to optimized conditions

Required standard

Key: 1 = Reactive, 3 = Stable running, 5 = Optimized intervention

FIGURE 6.19
Operational resilience assessment (case study).

the operational competencies needed to support this and lock in the gains made. The ticks show the assessment of current capabilities and standards during the project.

This assessment indicates that to deliver the full potential of the investment, this organization must include actions within the scope of the project to deal with weaknesses in

- The standardization of operating and asset care practices
- Problem-solving processes
- The visual control of normal/stable operating conditions
- Actions to trap and eliminate the causes of defects
- Spares management

6.5.2 Best-Practice Design Books

EEM best-practice design books are used to capture lessons learned and codify them as design goals, guidelines, and standards.

The structure of the best-practice design book will vary from company to company. These are generally organized into chapters and sections, one chapter per design module.

It is important to define these chapters by function rather than by the type of mechanism or vendor name to future proof this important resource against technological changes that reduce the value of the knowledge contained within each chapter.

For example, the best-practice design book for a food manufacturer should organized into chapters such as internal transportation rather than specific vendor equipment, automated guided vehicles (AGVs), or conveying. AGV and conveying options would be a section within that chapter.

Organizations that do this well distribute ownership of each chapter among engineering personnel. Their role is to provide a bridge between today's assets and future asset potential by

- Keeping the assigned chapters up to date
  - Monitoring the progress of focused improvement activities
  - Tracking changes in vendor offerings/technology
- Providing guidance to project teams working on projects to improve current assets or procure new ones

In those organizations, these chapter headings are also used to structure internal learning plans associated with skill development. Here, the allocation of ownership of a design book chapter to an engineer is treated as part of their learning path and as recognition of their advanced capability.

6.5.3 Compatibility of EEM with Other Design Tools

Despite the origins of EEM, the toolbox is sufficiently independent of TPM to be suitable for use alongside other continuous improvement processes such as Lean Manufacturing or Six Sigma. In addition, the discipline and structure provided by project management tools such as PRINCE2 can also be utilized as part of the project delivery process.

During initial EEM projects, those carrying out the project governance role should evaluate how to enhance current practices using EEM principles and techniques to ensure that

- New equipment specifications incorporate lessons learned on current assets over time.
- The full potential of existing assets is realized before large capital expenditure projects are considered.
- The new asset will be supported by effective operating and asset care practices from day one.

6.6 DEFINE

Project governance goals at the concept and high-level design steps are as follows:

- Identify the right scope.
- Involve the right people.
- Explain the business case.

6.6.1 Concept Stage Gate

Outputs from the concept step include:

- A project charter
- The relevant problem prevention data

- A concept assessment against the design goals
- Objective testing of the short list
- Targets for each of the six EEM target areas
- LCC model justification
- Project milestone plans
- A stage gate sign-off report plus documentation of lessons learned and feedback to design best-practice books

Audit outputs at this step assess how well the documents explain

- Critical process elements
- Design efficiency
- Operational conditions
- Process trade-offs
- Intrinsic reliability
- Operational reliability
- Other options considered
- Building ownership

6.6.2 High-Level Design Stage Gate

Outputs from the high-level design step include

- Team mobilization and an agreed team charter
- Evidence of options considered and rejected plus the basis for selecting the preferred option
- Outline specification, including
    - Actions to restore the basic conditions of existing equipment
    - Provisional layout, workflow changes, new asset footprint
    - Preferred equipment specifications
    - Design module functional specifications and checklists
- Financial justification and risk assessment/sensitivity analysis
- Invitation to tender (ITT) template, an evaluation of the long list of vendors, and the development of the vendor short list
    - For large projects, a report on the short list of vendors, including company data, company experience, QA/accreditations, EEM knowledge, and financial status
- A realistic and achievable outline project plan, including testing and inspection criteria for design modules

- A submitted/approved funding application
- Approval of the preferred option and readiness for submission
- Approval of the resources for the next step, including detailed design development

Audit these outputs to assess how well the document shows evidence of understanding of

- Potential vendor strengths and weaknesses
- Discounted process options
- LCC assessment
- Design module criticality/test definition
- Critical path and knowledge gaps
- Design standards set/required
- Buy-in from all parties to the preferred option

6.7 DESIGN

The project governance goals at the detailed design and prefab procurement steps are as follows:

- Select the right vendor.
- Tease out latent weaknesses.
- Design a robust solution.
- Enhance project value.
- Develop a realistic plan.

The precise regulations will depend on the country where the project is carried, but there will usually be a need to notify governmental health and safety authorities at this stage if the project

- Will last longer than 30 working days
- Has more than 20 operatives working simultaneously
- Exceeds 500 person days

There may also be a need to appoint a principle designer and principle contractor and to ensure that they comply with their duties.

Users of machinery may also have responsibilities such as the Provision and Use of Work Equipment Regulations (PUWER) 1998.

- Select and provide suitable work equipment, taking into account the working conditions and health and safety risks in the workplace.
- Ensure that it is used correctly.
- Keep it maintained in a safe condition.

When buying new equipment (including machinery), users may also be required to check it complies with all relevant supply laws—for example, that it is

- CE marked
- Supplied with a declaration of conformity and user instructions in English
- Free from any obvious defect (such as missing or damaged guards)

There may also be other statutory duties such as maintaining and inspecting the equipment to ensure it remains safe.

6.7.1 Detailed Design

Outputs from the detailed design step include

- Team mobilization and an agreed team charter; tenders issued to short-listed vendors
- Evaluation of each vendor ITT response and the identification of a recommended vendor
- Project visioning/vendor induction
- Confirmation of the project's desired outcomes, goals, strategies, and actions, and issues to deal with to secure the project's goals
- Identification of who does what and what is needed from site, the project team, and the vendors
- Confirmation of project risks identified and tactics to mitigate them
- Detailed budget and LCC forecasts for each design module
- Realistic and achievable detailed project plan
- An agreed specification and document management protocol covering ownership, control, and the management of agreed changes to the specification

Audit these outputs to assess how well the document shows evidence of actions to

- Assure all design targets can be met
- Minimize the risk of human error/intervention (operator and maintenance)
- Identify LCC/hidden-loss targets
- Define visual management/indicators
- Define tooling, cleaning, and maintenance aids

6.7.2 Prefab Procurement

Outputs from this step can include

- Production and maintenance working methods and underpinning systems, including spares, production materials, tooling, bills of materials, CMMS systems, planning processes, and QA tests
  - Includes plans for draft working methods to be refined alongside installation and commissioning activities. This is also aligned with the witnessed inspection program to assure that working methods support the delivery of stability and resilience goals.
- The induction into and management of installation protocols
- Installation plan development and mobilization; may be the responsibility of the main contractor
- A plan to support the progress management of installation and commissioning activities
- Problem prevention at key stages of the manufacturing/construction process; a manufacturing inspection report
- Witnessed inspection reports, including sign-off of the factory acceptance test (FAT)

Audit these outputs to assess how well the document shows evidence of actions to achieve

- Vendor communication/constraint feedback
- Quality audit criteria
- Detailed project planning
- Layout design
- Best-practice definition
- Training material
- Ownership development

6.8 REFINE AND IMPROVE

The project governance goals at the installation and commissioning steps are as follows:

- Prevent problems.
- Develop competence.
- Transfer learning.

Installation and commissioning best practice is characterized by

- Extensive integrated planning in detail
- The use of a single plan that defines
    - Contingency plans (i.e., when to continue and when to stop, take stock, and develop a new plan)
    - The critical path
    - The glide path to flawless operation
- A structured witnessed inspection program to capture and resolve latent weaknesses and control engineering change as a routine activity
- The systematic development of new competencies as part of the installation and commissioning process
- The capture of lessons learned in best-practice design books, including
    - The stepwise evolution of operational design
    - Updated standards and processes
    - Relevant company team and vendor knowledge

The biggest risk to flawless operation from day one is human error. Time spent refining working methods to reduce error and defect risk will make it easier to train in core competencies. Actions to define optimum conditions provide the learning process to develop intermediate and specialist skill levels. This learning process is accelerated by being carried out alongside vendor installation and commissioning personnel.

6.8.1 Installation

Outputs from this step can include

- Team mobilization and an agreed team charter
- Phased testing of equipment functionality as part of the witnesses inspection program
- A glide path plan to flawless operation, including
    - Activities to finalize the detailed design of the workplace

  - Refined working methods alongside testing and commissioning activities
  - The coordination of resources and activities to identify delays/early completions and compliance with safety and environmental practices
  - The tracking of installation performance and readiness to proceed to the next step
- Approved/refined commissioning plans and a glide path to flawless operation

Audit these outputs for evidence of the use of

- Installation feedback
- Workplace organization
- OEE and continuous improvement targets
- Pilot runs to refine best practice
- Objective problem-solving/optimization
- Breakdown briefings/communication
- Visual progress information
- Teamwork across all groups involved

6.8.2 Commissioning

Outputs from this step can include

- Team mobilization and an agreed team charter
- A list of tests carried out/problems addressed and actions taken
- A visual schedule showing the current status and next steps
- Captured installation weaknesses
- Evidence of competence in skills and underpinning knowledge to run and manage the new operation
- Individual test/problem-solving reports including details of the protocol used and actions taken following a successful/unsuccessful outcome
- Confirmation of the flawless operation date
- Handover of the latest drawings and draft manuals

Audit these outputs for evidence of the achievement of

- Design target levels
- Flawless operation
- Standardized best practice
- A definition of optimum conditions
- Up-to-date technical documentation

6.8.3 Post Day One: Stabilizing

A key project governance challenge at this stage is avoiding the risk that new equipment is allowed to deteriorate post day one. In the busy world of operations, the impact of such deterioration can remain hidden for some time.

Guide the transition of the project focus from achieving stable performance to systematically improving process precision and increasing the time between interventions.

The outcome directly impacts on profitability through increased productivity, improved material yield, and the ability to handle cheaper materials without losing output consistency. It also reduces overhead costs by increasing capacity, extending component life, and improving flexibility to shifts in demand.

After completion of the site acceptance test (SAT) and before the end of the warranty period, hold a lessons learned review on-site with the entire project team to

- Obtain honest, objective, and constructive feedback on the effectiveness of the project process
- Recommend changes that will improve the project value
- Support transfer of lessons learned to future project teams

6.9 CHAPTER SUMMARY

This chapter provides the following insights.

- How EEM support improved strategic control.
- Support for the leadership challenge of raising capability by providing the opportunity to develop those taking part in projects and those who backfill their positions.
- The benefits of closer working relationships with vendors.
- Progress is measured through changes in outlook.
- EEM is best integrated with a continuous improvement process to improve current asset performance and raise standards to support higher levels of technical stability and operational resilience.

Key learning points include the following:

- Develop a site EEM master plan defining commercial, operational, and technological challenges.
- Align these with site-strategic goals, guiding capital investment priorities, and setting the context for the scope of capital projects.
- Develop design guidelines, technical standards, and checklists to codify experience to guide design decisions and project management delivery.
- Create LCC models to identify cost drivers and support option evaluation.
- Manage the path to flawless operation.
    - Installation and commissioning
    - Developing operational capability
- Use the stabilization step to define the road map to optimum conditions. Make this the launchpad to better-than-new performance.

7

Implementing EEM

This chapter covers the implementation of EEM principles and techniques. The chapter also covers how to lock in the gains from EEM by integrating the approach to projects into the routine management process.

As many capital projects originate from the introduction of new products or services, the biggest gains from EEM are achieved when the principles are applied at the product design stage. This is covered in more detail in Chapter 8, "Early Product Management." The combination of early equipment and product management is known as *early management*.

The implementation plan uses the same steps as the EEM road map itself.

1. Define: Understand the need to improve.
   - Review the current capital project processes and identify strengths and priorities for action.
2. Design: Create an EEM policy.
   - Use a pilot project(s) to improve the current processes, define EEM standards, and deal with gaps in related business processes.
3. Refine: Make EEM systems efficient.
   - Train managers, design engineers, project managers, and CI facilitators in lessons learned from the pilot.
4. Improve: Use EEM extensively.
   - Capture lessons learned from each project and use to update the design standards. Reinforce EEM policy by training each new project delivery team, including key stakeholders.

The EEM implementation milestones are set out in Figure 7.1. This includes the planning of a pilot project to build on existing good practices and identify policy gaps.

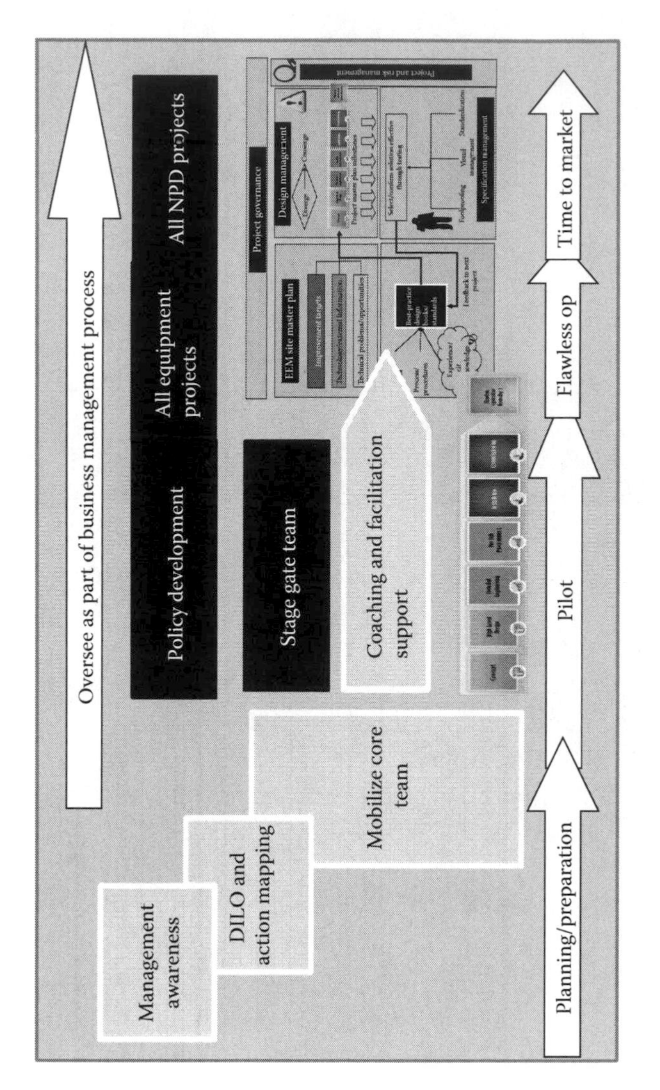

FIGURE 7.1
EEM implementation milestones.

The following sections describe how one multisite organization carried out their EEM implementation. This began with the need to improve the chances of success of an existing project at or near to the *high-level design* (HLD)/funding approval stage gate.

It is important that pilot projects do not delay the project delivery time line, but even if time were available, experience has shown that there is little to be gained in returning to an earlier project step. Nevertheless, always confirm the project brief and forward program by applying the HLD tasks even if the project has progressed past this point. Once the HLD tasks are completed, the forward plan should continue from the current project step.

7.1 DEFINE

7.1.1 Planning and Preparation

The planning and preparation step began with an assessment of current practices against EEM audit criteria to complete an EEM diagnostic. The assessment was carried out by a cross-section of experienced project members using the audit criteria set out in Chapter 6, "Project Governance," Sections 6.6 through 6.8. The assessment profile (using a 1–5 scale where 1 = little, 3 = acceptable, and 5 = outstanding) is shown in the Figure 7.2.

The shape of the spider chart profile highlights areas of strength and weakness. In addition, the scores for steps 1 through 3 measure how well the organization specified equipment and the scores for steps 4 through 6 measure the organization's project delivery capabilities. The scores shown in Table 7.1 indicate that the company is better at project delivery than it is at equipment specification.

The assessment of each step is carried out using a 1–5 scale. A score of 1 indicates a policy gap, 2 indicates that some processes are in place but they are not consistently applied, 3 indicates that a robust process is in place and is working, 4 indicates a process that has been modified to achieve a measurable improvement, and 5 indicates a process with a significant track record of improvement.

This shows how the organization had some useful practices in place around installation and commissioning but that these were not consistently applied. As with any change, it takes less effort to refine an existing, accepted approach than to start with a completely clean sheet of paper.

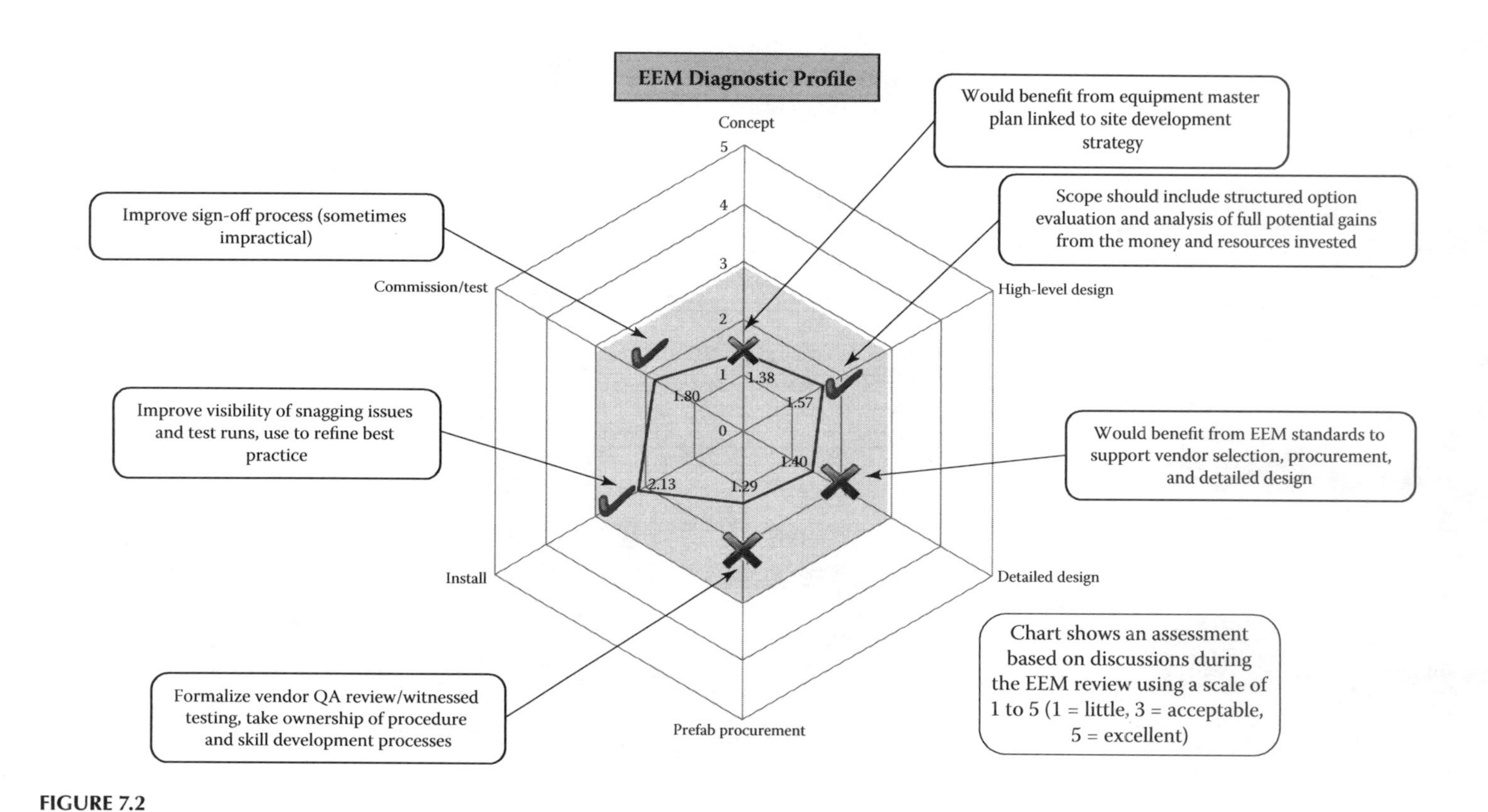

FIGURE 7.2
EEM process gap assessment (case study).

TABLE 7.1

EEM Diagnostic Assessment

	EEM Step	Assessment	% of Max.
1	Concept	1.38	28%
2	HLD	1.57	31%
3	Detailed design	1.40	28%
	Subtotal planning		29%
4	Prefab procurement	1.29	26%
5	Install	2.13	43%
6	Commission/test	1.80	36%
	Subtotal implementation		35%

The new process would only have to overcome the inertia that resulted in the weaknesses shown.

In particular, these weaknesses were symptomatic of an organization that had limited interaction between operations personnel and vendors prior to day-one production. It was also symptomatic of an organization that operated in functional silos. To deal with these weaknesses, the organization needed to improve its approach to cross-functional collaboration.

In addition, the organization relied heavily on the vendor experience, but as discussed in previous chapters, vendors can only directly influence around 50% of the factors that contribute to project success. The result was a capital project process that could be likened to a white-knuckle roller coaster ride and an asset pool of equipment with a mix of adequate and lackluster performance.

An EEM implementation plan was designed to develop a balanced capability for each step before progressing to the next level. In this case, that meant formalizing a policy to achieve a score of 2 in steps 1 (concept), 3 (detailed design), and 4 (prefab procurement).

The pilot program was designed to address the areas of weaknesses and firm up areas of strength.

- Mobilizing a cross-functional EEM core team to codify operational knowledge into design guidelines and checklists to support the detailed design and project delivery processes
- Close collaboration with vendors to improve insight of how to get the most value out of the new asset
- Support from internal and external specialists to work with the core and stage gate teams to help them develop improved ways of working and raise internal capabilities to apply them

In this way, in addition to improving the project performance, the deliverables of the pilot were policy recommendations and a program to implement them fully. Two projects were selected: the original project, currently at the HLD step, and an additional project at a second site to work on processes for the concept step.

7.1.2 Management Awareness

The outputs from the diagnostic and pilot planning activity were used as part of a management awareness session to

- Raise awareness of EEM principles and techniques
- Explain the pilot process
- Agree goals and success criteria for the pilot
- Agree the process for developing and deploying lessons learned as policy guidelines

7.1.3 Mobilization

Once approved by the management team, the outputs from the planning activity were then used within cross-functional teamwork sessions to

- Raise awareness of the EEM core team and project sponsors
- Add detail to the EEM milestone plan steps for the pilot project
- Set out an action plan to progress to the next EEM milestone

The following agenda sets out the three-day program used.

Section	Content
	Day 1
EEM introduction	What goes wrong, EEM goals and principles, overview of EEM steps
EEM standards	Understanding EEM standards, EEM criticality assessments, and the *day in the life of* (DILO) review process
Concept and HLD	Compare proposed design with EEM design goals to assess design stability and resilience and value engineering opportunities to improve added value
EEM milestone plan	EEM project organization, the stage gate review process, and audit criteria applied to future project steps

(*Continued*)

Section	Content
	Day 2
Detailed design and prefab procurement	Developing the forward design review process and installation planning steps
Installation and commissioning	Planning the steps to flawless operation and beyond
	Day 3
Project governance	The stage gate review process and tasks to complete to meet the next EEM milestone
Action planning	Collate plans from previous sessions into a pilot milestone plan and detailed actions to the next EEM milestone
Feedback	Feedback of next steps

Actions agreed following these working sessions included

- Confirmation of the project workstream goals, timetables, and resource plans developed during the sessions
- The development of a vendor questionnaire to assess vendor alignment with EEM goals
- The completion of a review of each design module to identify EEM design guidelines and checklists for each module
- An assessment of the current operational processes to identify opportunities to reduce levels of intervention and quality assurance (QA) testing
- The development of a draft layout showing the equipment footprint and material flows, plus an outline of ways of working to formalize the HLD specification
- The development of detailed plans for the detailed design step and outline plans for each EEM step
- Completion of the HLD stage gate review

7.2 DESIGN

7.2.1 Pilot

The EEM core team started their EEM journey at the HLD step and followed the tasks for each EEM step, as set out in Chapter 2, "The EEM Road Map."

The DILO activities and module review processes from the two pilot projects generated the design guidelines. This highlighted a number of

TABLE 7.2

Module Review Output Summary (Example)

Issues Raised		Action Categories	
Module scope	2	In spec	81
Safety	1	Out of spec	12
Reliability	59	Added to spec	40
Operability	52	Develop concept	9
Maintainability	19	Future checklist point	20
Customer value	12		
LCCs	17		
Total	162	Total	162

design issues to be addressed, including nine concept gaps and 40 additions to the specification (Table 7.2).

These outputs were also used to create design guidelines for each of the EEM goals (Figure 7.3).

These EEM design guidelines were then used to support the vendor selection, detailed design, and witnessed inspection processes.

The example in Figure 7.4 is a summary of the assessment of vendor options using the EEM guidelines. This involved

- Visits to site by a short list of four suitable vendors to brief them about the EEM process and requirements contained within the design guidelines.
- Each vendor completing a self-assessment questionnaire, wherein they rated the fit of their proposed solution against each of the guideline headings using a 1–5 scale (1 = not met, 3 = met, 5 = met at multiple reference sites). Vendors were asked to provide evidence of their assessment.
- A review of vendor assessments by the EEM core team to calibrate the scores against the evidence provided.

Although vendor 1 scored the highest overall, vendor 2 scored higher in terms of operability. Follow-up discussions/meetings were held with the top three vendors to clarify their understanding and discuss the perceived strengths and weaknesses of their proposal. Ultimately, the contract was won by vendor 1. Although the EEM assessment was only one of the factors involved in the decision, in addition to providing the best value for money, the vendor also addressed the maintainability weaknesses identified so that their offer scored highest in all sectors.

Design Guideline Headings						
	Safety	Reliability	Operability	Maintainability	Customer value	LCC
1	Access	Changeover and settings	Operator access egress and lighting	Maintainer access egress and lighting	Specification	Communications
2	Product safety/QA	Material usage	Line start-up and run outs	Engineering	Verification	Machine performance
3	Change parts	CIP/santization	Changeover	Environment	Installation hygiene	Preform quality
4	Utility hazards	Work place organization	Material logistics	Ergonomics	Hygiene	Process control
5	Lasers	QC checks	Operator intervention	Hand tools	QA tests	QA inspection
6	Risk of injury from heat during set up or clearing jams	Protection	Standardize/simplify best practice	Inspection	Defect source	Cause of defect/loss
7	Motors	Jams	Raw material storage	Prevention	Defect detection	Traceability
8	Security of local isolation state	Process control	Work station design	Repair	Finished goods material flow	Compressed air
9	Ease of problem detection at a distance		Workplace organization	Routine servicing	Flexibility	Effluent
10	Segregation of fork lift trucks		Contamination	Cleaning		Electricity
11				Spares		Energy recycling/saving
12						Productivity
13						Maintenance cost
14						Engineering spares
15						Material yield

FIGURE 7.3
EEM design guideline headings (case study).

Vendor 1			
Intrinsic safety	3.59	3.51	3.49
Intrinsic reliability	3.44		
Operability	3.70	3.51	
Maintainability	3.32		
Customer value	3.56	3.44	
LCC	3.33		

Vendor 2			
Intrinsic safety	3.32	3.38	3.34
Intrinsic reliability	3.44		
Operability	3.45	3.52	
Maintainability	3.59		
Customer value	3.11	3.12	
LCC	3.13		

Vendor 3			
Intrinsic safety	3.41	3.36	3.30
Intrinsic reliability	3.31		
Operability	3.55	3.46	
Maintainability	3.36		
Customer value	2.94	3.07	
LCC	3.20		

Vendor 4			
Intrinsic safety	2.86	2.93	2.98
Intrinsic reliability	3.00		
Operability	2.85	3.02	
Maintainability	3.18		
Customer value	2.89	2.98	
LCC	3.07		

FIGURE 7.4
Vendor option assessment.

The pilot process provided practical, hands-on experience of how the current reality compared with the benchmarks of EEM best practice. Through this, the management team were able to recognize and deal with weaknesses, including

- Their equipment management vision (what "good" looks like) and mission (how to get there)
- Key performance indicators to track progress, prioritize actions, and set improvement targets
  - The definition of and links to business drivers
  - Tracking and reporting mechanisms
- Implementation planning, organization, and control
  - The identification of EEM business sponsors and subject matter experts
  - The policy deployment process
  - The training and support program and resources
  - Projects within the scope of the roll-out cascade

    - Site master plan development
    - The review and action meeting schedule
- Site equipment master plan development as part of strategic planning
- Best-practice design book development and the feedback process

7.2.2 Policy Development

For this organization, the pilot identified gaps and provided insight to support policy development in the following areas:

- The road map and stage gate process
    - How to manage the evolution of design specification, delivery partners, and project delivery
- Project roles and accountabilities
    - Defined cross-functional project team accountabilities
- Incremental and step-out product/service development processes
    - The process for translating customer step-out value, adding features into product and service design
- LCC modeling and management
    - How to develop it, and where and when to use it
- Concept development
    - Front-end evaluation and definition using design goals, guidelines, and technical standards
- Equipment design guideline standards to support procurement, detailed design, and witnessed inspection
    - Definitions of operability, maintainability, intrinsic reliability, and intrinsic safety
- The collation of current knowledge
    - Processes to collate explicit and tacit knowledge and identify knowledge gaps
- Managing technical knowledge gaps
    - Defining and deal with areas of risk

These points were incorporated into policy guidelines and a deployment process was developed for the five EEM subsystems discussed in Chapters 3 through 6.

This company used their existing policy deployment process to cascade EEM policy across the other sites. This was supported by the use of a number of templates (Figure 7.5). Pilot project examples were used to populate the templates as visual standards to support the policy roll-out.

	Purpose	Created at	Also used for
1. Module definition	Provide structure for design development Collate information and issues/concerns from module review	High-level design	Procurement, detailed design, and development of operational working methods
2. Objective testing	Support comparison of option strengths and weaknesses	Concept	Decision support tool for all stage gates
3. EEM milestone plan	Set project key steps and timings	High-level design	Project governance tool for all stage gates
4. Quality plan	Define the sequence of outcomes needed to assure the success of the project	High-level design	Updated at detailed design with witnessed inspection schedule and all assurance testing to SAT
5. Target setting	Clarify areas of defined and potential added value	High-level design	Once the final option has been selected, targets should be updated to take account of additional gains/added value to be achieved from the project
6. Vendor selection questionnaire	Support the evaluation of vendor offerings against EEM design goals and operational standards	High-level design	Used as part of evaluation of vendors and during detailed design discussions to assess design effectiveness and tease out latent design weaknesses
7. Witnessed inspection schedule	Summarize the project quality assurance tests/checks to confirm progress toward project goals	Detailed design	Used/updated at all steps to site acceptance testing

FIGURE 7.5
EEM template list (case study).

7.3 REFINE

The implementation program followed on from the pilot, as set out in Table 7.3. Each site was tasked with completing an initial project to introduce the site to EEM and allow the opportunity to fine-tune policies to their specific asset pool and product range.

The roll-out cascade was managed at site level, with implementation benchmarks for EEM policies being added to the company's continuous

TABLE 7.3

EEM Implementation Plan

	Initial Project/ Pilot Lessons	Roll-Out Actions	Locking in the Gains
1. Agree what needs to change and why.	Review the existing design process and select initial projects and personnel.	Involve key personnel in leading their first early management project. Define cross-project improvement targets.	Use case studies as part of the induction process for new team members and project sponsors. All new projects use EEM.
2. Mobilize EEM project.	Mobilize initial project team(s). Develop initial project standards and checklists.	Launch teams and facilitate cross-project learning.	Establish early management facilitators to multiply benefits from EEM projects.
3. Collate current operational knowledge.	Use manual simulation and analysis of criticality document weaknesses. Systematically record problems/ improvements.	Focused improvement teams resolve chronic losses and structure the knowledge base. Individuals are identified as best-practice design book guides.	Feedback breakdown review outputs and single-point lessons to best-practice design book guides. Establish the *one idea, twenty opportunities* approach.
4. Use projects to develop new competencies.	Establish standards, best-practice design book use, and the feedback and stage gate review process.	Establish EEM site master plan and deploy. Review as part of management's quarterly performance management process.	Reinforce ways of working. Record and apply focused improvement to the EEM process.

improvement audit so that senior management could track progress at each site.

The implementation of EEM policy was also led by internal site project sponsors, project managers, and EEM subject matter experts as part of the pilot project delivery. Their tasks included the definition of EEM best-practice design books and updates to support current and future capital projects.

The site plans applied a three-phase cascade. Phase 1 involved the delivery of a first project by each site and each project manager. Progress measures included an assessment of demonstrated competence. Each site was supported by external specialists to provide training and coaching support where needed. This began with foundation training, as set out in Table 7.4.

Support was also linked to the learning plans set out in Section 7.4. This included

- An initial readiness review
- EEM core team and stage gate team coaching for the following project governance phases:
    - Define, including the capture of tacit knowledge as design standards/guidelines and option evaluation processes
    - Design, including vendor selection and detailed design/planning
    - Refine, including the application of witnessed inspection, operational change management, and coordination of the glide path to flawless operation

TABLE 7.4

EEM Foundation Training Plan

Training Event	Duration	Content	For
Executive workshop	1/2 day	EEM overview plus senior management role	Key stakeholders
Foundation learning	3 days	Foundation learning and pilot mobilization	EEM core team and project sponsors
EEM stage gate awareness	1 day	EEM project governance, master plan, and stage gates	EEM project sponsors and facilitators
EEM core team mobilization at other pilots	1 day	Introduction to EEM and development of EEM milestone plan	EEM core team (1 day for each step)
General awareness	2–4 h	Introduction to EEM plus project roles	Those involved in part of the EEM program

Once a site was able to demonstrate its capability to successfully pass through at least three stage gates with its initial project, the second phase of the roll-out cascade could begin to apply EEM to all new projects. During the final phase of the roll-out cascade, the use of EEM was widened to include new product and service development.

7.4 IMPROVE

As with the development of any skill, capability and confidence are achieved through application. Table 7.5 provides an overview of the training and application experience used to confirm competence and set out the learning plan for each key stakeholder.

7.4.1 EEM Subject Matter Expert Learning Pathways

At the center of the site EEM implementation process is the EEM subject matter expert. Their development is supported by learning plans for

- Design and performance management
- Specification and LCC management
- Project and risk management
- EEM principles and techniques

These are set out in Tables 7.6 through 7.9, which also define core, intermediate, and specialist topics. As shown in the preceding table, this journey begins with a three-day practical EEM workshop to raise awareness of core competencies and support the development of personal development plans.

7.5 SPEEDING UP TIME TO MARKET

After establishing a robust project delivery process, the focus shifts to improving the five EEM subsystems introduced in Chapter 2. Whereas the delivery of flawless operation from day one is evidence that each system has achieved an acceptable score of 3 on our 1–5 scale, some work may

TABLE 7.5

EEM Stakeholder Learning Plan

	Assessment Basis (1 = Limited, 3 = Acceptable, 5 = Master)				
	1	2	3	4	5
	Aware	**Can Apply with Support**	**Core Competency**	**Intermediate Competency**	**Specialist Competency**
EEM project sponsor/ pillar champion	Attended EEM management awareness session	Completed project mobilization and two stage gate reviews	Planned future EEM step activities and achieved EEM audit score of 3	Led EEM step process for that role and achieved EEM audit score of 3	Led EEM projects in that role for all steps and achieved EEM audit score of 3 or more
EEM core team member	Attended general awareness training	Completed two core team steps			
EEM facilitator/ project manager	Completed EEM 3-day specialist workshop	Completed EEM team mobilization and two stage gate reviews			

TABLE 7.6

SME Design Management Learning Plan

Design Management				
Title	**Topic**	**Content**	**Level**	**Project No./Date Achieved**
Defining equipment performance drivers	Defining the modules, assessing understanding, highlighting potential areas of weakness, understanding losses and countermeasures, understanding LCCs	Module definition, criticality assessment, overall equipment effectiveness, hidden losses, flow, setting targets, module review, understanding losses, defining commissioning criteria	Core	
Using standards and checklists	EEM design goals, guidelines, and standards	Using EEM standards to guide decisions when defining options and working with vendors	Core	
Developing operational standards and checklists	How to codify experience to guide design decisions and project management delivery	DILO cluster issues under design goals (top 10 topics), developing acceptable and optimum criteria	Intermediate	
LCC cost model	Creating LCC models to identify cost drivers and support option evaluation	Capital and operational cost streams with parameters for throughput, product mix, and effectiveness levels	Specialist	

TABLE 7.7

SME Specification Management Learning Plan

Specification Management				
Title	**Topic**	**Content**	**Level**	**Project No./Date Achieved**
Making choices, refining design	Defining modules/functions, generating and evaluating options, selecting and refining the preferred option	Value engineering, objective testing, LCC models	Core	
Selecting and working with vendors	How to assess offerings, identify a short list, and select a preferred vendor in a way that builds collaboration and shared goals	ITT review, vendor induction, detailed design workshops, and project planning	Core	
Developing a URS	Confirming the brief, thinking through the URS/ITT content	Aim, principles, ways of working, systems, procedures	Intermediate	
Developing a site EEM master plan	Defining commercial, operations, and technology challenges linked to site's strategic goals to guide capital investment priorities and individual project scope	Defining commercial, operations, and technology changes and challenges; site investment scope, tactics/sequence, and equipment management goals	Specialist	

TABLE 7.8

SME Project Management Learning Plan

Project Manager				
Title	**Topic**	**Content**	**Level**	**Project No./Date Achieved**
Developing project plans	Milestone planning, setting exit criteria to define project quality plans	Quality plan timing and the completion of internal documentation (e.g., the capex process)	Core	
Using project quality plans	Carrying out witnessed inspection	Test criteria	Core	
EEM links with other business processes	Capex, PMS, QA , documentation, risk management, legislation	Understanding where EEM fits in	Core	
Developing collaborative plans	Action mapping for project scope definition and vendor induction	Facilitating cross-company events to agree a consolidated project strategy and supporting workstreams	Intermediate	
Project governance	How to design, mobilize, and support outstanding capital project teams	COT targets, RACI steps, and stage gates	Specialist	

TABLE 7.9

SME EEM Principles and Techniques Learning Plan

EEM Principles and Techniques				
Title	**Topic**	**Content**	**Level**	**Project No./Date Achieved**
The EEM Route	The decisions at each stage, the processes to be sure we don't miss things, and the processes to prevent issues from being transferred to the next step	Milestone plan, tasks, RACI	Core	
Get the right design	Get the scope right, concept design, and HLD		Core	
Get the design right	DD and prefab procurement	Witnessed inspection, designing the training cascade	Core	
Manage the equipment delivery	Installing and commissioning, developing an understanding and training process	Delivering the training cascade witnessed inspection	Core	
Manage the operations changes	Developing capability	Training cascade	Core	
EEM Process	Behaviors, collaboration, using working assumptions	Facilitation	Core	

be necessary to be confident of sustaining that score. Lock in the gains by refining each EEM sub system to progress from 3 to 4 and 4 to 5 on the maturity scales. The gains will be felt in increased innovation, collaboration and policy deployment. Once the momentum is lost, it can be difficult to regain it.

The gains to be made by improving EEM performance include a systematic reduction in time to market and the resources needed to deliver the project goals.

7.5.1 Design and Performance Management

Purpose: The systematic evaluation of options and the selection of a preferred design approach to satisfy each EEM decision step exit criteria (Table 7.10)

Capital projects are often mobilized to support a new product or service. Companies that do this well work closely with suppliers and customers to optimize design features and improve supply chain responsiveness. Improving design and performance systems to simplify the process of developing suitable alliance relationships will make it easier to create new value and competitive advantage.

TABLE 7.10

Design Management Route to Excellence

1. Weak	3. Acceptable	5. Excellent
Informal or missing design development steps. Heavy reliance on vendor to deliver what is needed. Emphasis on technical performance. Little or no front-line involvement pre-installation.	Design guidelines used to raise vendor understanding of operational needs. Cross-functional teams select and refine design options from concept step.	History of improved design standards leading to industry-leading performance.

7.5.2 Specification and LCC Management

Purpose: To refine the preferred design at each step to minimize LCCs and enhance project value (Table 7.11)

As the specification evolves through the steps of concept to detailed design, the specification process guides decision making and captures lessons learned. Support this through the capture of knowledge about how to achieve optimum conditions, reduce planned intervention levels and

TABLE 7.11

Specification Management Route to Excellence

1. Weak	3. Acceptable	5. Excellent
Relies on contractor documentation. Limited attempts to tease out latent design weaknesses/refine.	Specification process used to reduce LCC and enhance project value.	Significant history of reduced LCCs and increased project value.

sources of quality defects. This in turn will improve flexibility and the ability to produce more closely to customer demand cycles. This includes work on improving product design for manufacture.

7.5.3 Project and Risk Management Processes

Purpose: The planning, organization, and control of resources to complete EEM step activities, manage risks, prevent problems from being transferred to future steps, and deliver project goals (Table 7.12)

Progress here depends on

- The integration of EEM and early product management activities
- The alignment of investment strategies as part of the equipment management master plan
- The development of management-standard work for projects to support self-managed high-performance project teams

TABLE 7.12

Project and Risk Management Route to Excellence

1. Weak	3. Acceptable	5. Excellent
Uses contractor project management processes. Relies on contractual clauses to manage risk.	Formal project templates and checklists used with stage gate reviews to trap oversights and opportunities to enhance project value.	Systematic reductions in resources needed to deliver flawless operation.

7.5.4 Project Governance

Purpose: Project mobilization, the stage gate review process, project close-down and the capture/transfer of lessons learned (Table 7.13)

Progress here depends on

- Establishing the link between strategic intent and strategy deployment as a cross-functional, team-based project delivery process
- Succession-planning processes to release experienced personnel to support projects and support the development of those backfilling roles
- Adopting innovation-friendly processes to encourage horizontal (cross-functional) learning, engaging and empowering the company team with the delivery of strategic goals

TABLE 7.13

Project Government Route to Excellence

1. Weak	3. Acceptable	5. Excellent
Senior management governance limited to financial oversight and ad hoc involvement to deal with problems as they occur.	Formal stage gate review process in place for each EEM step. Supported by audit/coaching to develop cross-functional teamwork and project delivery capabilities.	EEM stage gate process used for all projects, including annual shutdowns, new products and services, and IT systems. Projects seen as a vehicle to improve cross-functional collaboration and raise site capability.

7.5.5 Best-Practice Design Book Processes

Purpose: The codification of knowledge, technical data, and improvements into design goals, guidelines, and standards to support the delivery of low LCC design and the delivery of flawless operation from production day one (Table 7.14)

TABLE 7.14

Best Practice Design Book Route to Excellence

1. Weak	3. Acceptable	5. Excellent
Some technical standards defined. No change control, not routinely updated to reflect improvements made.	EEM best-practice design books include supporting design guidelines for all six EEM goals. Design guidelines used to identify weak components in current assets and guide investment decisions.	Accountability for design guidelines and standards update deployed to capture feedback from front-line improvement activities and changes in technology.

Typically, organizations at the start of their EEM journey have to invest project resources into developing this information. Furthermore, despite the use of EEM processes such as DILO and module reviews, it normally takes a couple of projects to establish a reliable set of design guidelines and standards because tacit knowledge is difficult to explain, write down, or transfer. It is also important to keep them up to date to reflect the current state of learning as problems are solved or effectiveness is improved.

Organizations that do this well split the task into functional processes and allocate the role of maintaining each best-practice design book section. The role of those allocated this task is to keep abreast of changes in technology and best practice and to be a point of contact on that topic. They become a bridge between today's assets and future capability. In some leading organizations, this role is used as a reward for those who show a passion for sharing knowledge and learning.

7.6 CHAPTER SUMMARY

In summary, this chapter covered the following topics:

- EEM implementation steps
- Implementation milestones
- Case studies
  - Diagnostics
  - Management awareness
  - Mobilization
  - Pilot projects
  - Policy development
  - Roll-out cascades
  - EEM learning plans and competency assessments
- Speeding up the time to market

Key learning points include the following:

- Measure organizational strengths and weaknesses using EEM audit criteria.
- Use practical pilot projects to learn how to apply EEM principles and techniques.

- Create an EEM learning plan for key stakeholders and project team members.
- Support their efforts to integrate the way projects are run into the business routine.
- The success of EEM depends on knowledge sharing and collaborative behaviors.

8

Early Product Management

Often, it is the desire to deliver a new product or service that triggers the decision to invest in new equipment or upgrade existing assets. That means that decisions made during the product or service design process have a major impact on the successful return on investment (ROI). Early product management (EPM) applies the principles of EEM to new product development (NPD) and new product introduction functions. In this chapter you will see that four of the five EEM subsystems (design, specification, project management, and project governance) can be extended to support EPM.

The scope of EPM covers two types of product/service development.

- Incremental improvements of existing products
- Step-out development of new products

8.1 WHY DO WE NEED EPM?

Back in 1990, the FDA reported that over 40% of product recalls over a six-year period were due to weaknesses that could have been dealt with during the design process. Since then, there has been little evidence to suggest that things are getting better.

It can be difficult to pinpoint the precise reasons for failures or the opportunities for improvement. As in the case of equipment projects, product development projects involve an iterative process of technical learning and discovery. Knowing whether one more or one less cycle of iteration would make a difference is impossible to measure. In addition, customers do not always know what they want, so asking them directly can be a frustrating process. Even when you do have a winning idea, the route to market can be equally frustrating.

Despite this, companies that have worked on improving product development processes have reduced development time by as much as 30%–50%—a prize worthy of consideration.*

8.1.1 What Is EPM?

The foundation for EPM success is good *knowledge flow* across commercial, operations, and technology functions. This includes processes to

- Gain a priceless insight into what customers really want
- Test and refine new ideas quickly with minimal effort
- Capture lessons learned and trap latent design defects at key stage gates

8.1.2 EPM Road Map

The EPM road map has three steps, as set out in Table 8.1.

This chapter covers the relationship between EEM and EPM and how they work together. Also covered are the three EPM steps, including project governance to avoid common pitfalls.

TABLE 8.1

EPM Steps

Step	Content
Shell	Clarification of the commercial case and potential routes to market, including operations process development and outline financial justification.
Shape	Evaluation of a short list of options to identify the preferred approach with a more detailed estimate of costs and benefits.
Scope	A detailed justification and process route and a route to market prior to formal sign-off of capital. This step could also include the production of commercial quantities using trial equipment to confirm the process route.

* Paul Adler et al. carried out a study into organizations with superior speed-to-market performance. *Harvard Review*, March 1996. Retrieved from https://hbr.org/1996/03/getting-the-most-out-of-your-product-development-process.

8.2 EPM STEP 1: SHELL

8.2.1 Design and Performance Management

The aim of this step is to generate innovative product and service concepts that create new value and advantages for customers.

The conditions for this to occur are

- A clear trigger for innovative thought (having a clear direction for what you want to achieve)
- A blend of outlooks (getting the right people in the room)
- Processes that consider a wide range of options and then converge on a preferred option

Although in theory innovative thinking should be unconstrained, in practice each NPD project should be part of an innovation stream that builds an organization's strategic capability. At its most successful, this will set the market agenda and keep competitors chasing your lead.

For that reason, innovation triggers should reflect the strategic drivers for your operation. For example, the innovation focus for organizations that provide *bespoke* products will be different from those with *fashion/brand* or *commodity product* business models (Table 8.2).

From the research carried out by Genrich Altshuller when developing the *theory of inventive problem-solving*, around 99% of innovative ideas occur by using existing mechanisms or processes in a novel way. The Apple iPhone and the Wright brothers' flying machine are classic examples of innovations as a result of applying things that exist in a way that offers new value and advantage.

TABLE 8.2

EPM Shell Step Tasks

Task	Purpose
Mobilize core team	Confirm the product business case. Develop a short list of options to evaluate. Set out the resource plan. Provide/allocate testing resources.
Refine product proposal	Develop commercial and technology propositions and outline the shape/scope proposals.
Develop route to market	Confirm commercial, operations, and technical approach/targets. Assess technical or commercial lag.
Shell sign-off	Agree the justification, priorities, and resource plan to develop plans to progress to the shape step. Add project to the capital program.

As a starting point, identify a desired customer profile, one that reflects your long-term strategic plans. Analyze the *total cost of ownership* for that customer using your product or services from procurement to disposal. Use this to gain insight into potential products and services and provide a trigger for innovation. Focus on how you could make their life easier/where new value could create advantage—for example, shelf-ready packaging, managed inventory, or preloaded software. Get close to strategic partners to help develop and refine the product.

8.2.2 Specification and Life Cycle Cost Management

Take the three best ideas and evaluate them against commercial, operational, and technological design standards to identify a preferred option. The design standards used at this stage should be defined to assess whether the concept is worthy of further investigation. Develop simple best-case/worst-case scenarios to assess risk and reward on a life cycle cost (LCC) basis.

8.2.3 Project and Risk Management

Risk assessment at this step should include steps to "know what you don't know" and to "confirm working assumptions." Paradoxically, putting in too much detail at this stage can unnecessarily consume resources and build ownership for weak product ideas. Avoid both risks by getting a mix of commercial, operational, and technological expertise involved in the decision-making process.

8.3 EPM STEP 2: SHAPE

The outputs from the *shell* step are some of the main drivers of capital investment. This in turn shapes the manufacturing assets/service procurement, which then shapes the operational and customer service capability.

The aim of the *shape* step is to select winning ideas and shape them into practical product and service offerings (Table 8.3).

8.3.1 Design and Performance Management Issues

At this stage, there is little quantitative data to support the evolution of outputs from the innovation step into a robust market offering. Develop a clear set of commercial, operational, and technical benchmarks/design

TABLE 8.3

EPM Shape Step Tasks

Task	Purpose
Mobilize core team	Confirm the status of options and the basis of the mid-step review.
Core team activity	Build and shape the project development network. Test and clarify the technical process. Confirm the project location. Identify facilities to confirm proof of concept. Select a preliminary product/service design. Design for the manufacture review. Plan steps to the scope stage gate.
Mid-step review	Confirm the selection process and preferred option. Review the preferred option, and red, amber, green issues.
Prepare shape proposal	Confirm the preferred option/proof of concept and priorities for actions to complete the plan to develop a proposal for the scope stage gate.
Shape sign-off	Agree the justification and resource plan to develop the proposal for the scope stage gate.

standards to support this task. What does "good" look like and how will it be achieved? The scope of these design standards should cover customer value, LCC, operability, maintainability, reliability, and safety.

Voice of the customer analysis is a great tool to increase understanding of customer value. Use this to add detail to the concept and to understand how customers compare options.

Research by MIT suggests that you can improve success at this stage by taking into account the broad market needs using the following tests:

- Watch what customers do. Are the current product/service features good enough to meet customer needs?
  - If yes, focus on new step-out products to disrupt competitor markets.
    - Test: Does the product/service give access to a new level of performance/value?
  - If no, focus on new features to improve market penetration.
    - Test: Does the product/service give access to currently unattractive market sectors or keep competitors out?

8.3.2 Specification and LCC Management Processes

Consider options to make it easier to manufacture/supply—for example:

- Minimize the number of parts or subassemblies.
- Develop a modular design using standard or multiuse parts.

- Avoid separate fasteners and adjustments.
- Design for tool-less top-down assembly with minimum handling.
- Simplify operations using repeatable and understood processes.
- Design for efficient and adequate testing and the easy analysis of failures.
- Rigorously assess value-added mechanisms and processes.

8.3.3 Project and Risk Management Processes

Risk assessment at this step should consider the levels of resources needed to deliver the next step. Too many projects running present a real barrier to progress.

Learn how to shelve/kill off ideas so that resources can be targeted at those projects with the most promise. When shelving projects, identify what it would take to achieve success criteria and revisit the shelf when those strategic triggers are met.

8.4 EPM STEP 3: SCOPE

The aims of this step are to obtain funding/resources, find suitable partners/vendors, and convert a robust market offering into a robust operation/supply specification (Table 8.4).

8.4.1 Design and Performance Management

As with the earlier steps, use innovation-friendly processes to add definition to the scope of the preferred product/service and route to market (diverge then converge). What is the best way to commercialize the idea: manufacture, outsource, license technology, acquisition, or joint venture?

Quantitative information is still scarce at this stage. Develop your organization's capability to develop and apply qualitative design standards (e.g., what does *ease of maintenance* look like?). This helps improve knowledge sharing between functions/organizations and encourages innovation.

Selecting the right supplier or equipment vendor may simply be a matter of building on the relationship developed during the earlier stages. Either way, organizations that do this well are those who recognize that the key to success is collaboration.

The skill at this step is to tease out latent design weaknesses and enhance project value. Use LCC analysis to target the main contributors to capital

TABLE 8.4

EPM Scope Step Tasks

Task	Purpose
Mobilize core team	Confirm brief with expanded team and support functions.
Core team activity	Confirm construction and operations configuration, confirm/refine linkages between cost and design criteria. Develop high-level design module briefs and standards for detailed design work. Review flow sheet operability/maintainability.
Prepare scope proposal	Integrate with high-level design EEM activity. Develop a short list of vendors. Program the release of personnel for the project delivery stage gates. Confirm capital costs and critical work programs.
Scope sign-off	Agree justification and resource plan to deliver the project. Appoint key personnel.

and operational costs, inject best practice for ease of use and ease of maintenance, and understand how to achieve intrinsically safe and reliable processes.

8.4.2 Specification and LCC Management

Get close to strategic suppliers/equipment vendors and leverage their knowledge of how to translate ideas into products and services. What will it take to achieve low LCCs (i.e., low capital investment and low operational costs)?

Don't be afraid to spend time visualizing ideas using sketches/diagrams and desktop/low-cost cardboard models. Don't be too quick to develop detailed drawings and expensive modeling. Detailed drawings will be easier to sign off once you are confident about how things will work together. Remember the old carpenters adage "Measure twice, cut once."

8.4.3 Project and Risk Management

Take care to identify and deal with knowledge gaps. Every NPD project will have new things to learn; make sure you spend time nailing down the few critical issues before firming up the details.

Work closely with vendors to develop a common vision and share ideas. Make sure that all parties are committed to delivering flawless operation from day one.

8.5 LINKS WITH EEM

The three steps of the EPM road map overlap with the first two steps of the EEM road map, as shown in Figure 8.1.

8.5.1 Project Governance

In addition to the project governance issues set out in Chapter 6, EPM presents a number of additional challenges.

Technology-led companies are often in the position of creating new markets. The project governance role here includes balancing the pace of technical development against the pace of commercial/market development. This can mean shelving further technical development at the current step until a commercial route to market has been developed or vice versa. Policy guidelines for these decisions will depend on the industry and market context. Figure 8.2 sets out policy guidelines used by a specialist technology business to measure the progress of product development and the allocation of resources. For example, the sanction of technology resources at the shell stage gate would only be made if the forward program included the commitment of a suitable strategic customer or alliance partner to support market-testing/proof-of-concept activities.

8.5.2 Design and Performance Management

As the specification is converted into physical operations, the key focus is the development of operational capability across all links in the supply chain.

Design performance should include consideration of how to manage pipeline filling, flexing to meet the demand profile once the product hits the streets and the direction of incremental improvement to stretch the product life cycle.

8.5.3 Specification and LCC Management

An important activity at this step is the assurance of the quality of construction and the provision of the physical operational aspects of the plan, source, make, deliver, and recycle process.

Again, don't be afraid to trial and test new aspects of the operation to trap problems before day-one production.

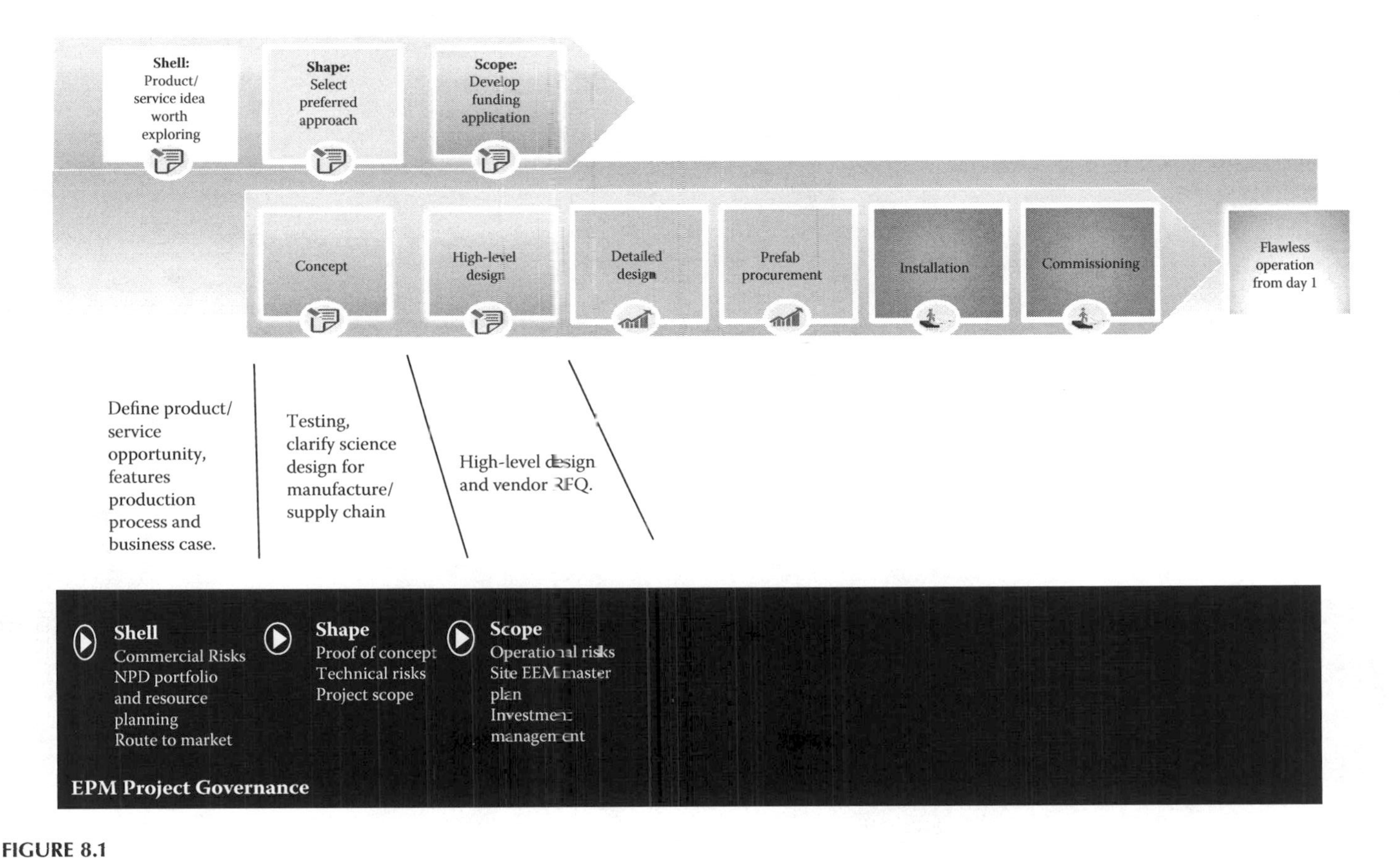

FIGURE 8.1
EPM stage gates.

Technology status

	Status	Characteristics	Evidence
SHELL	Defined features to master and potential gains to achieve customer/ competitiveness goals	Unfocused technical discussions about what might be possible	Defined target and science principles Understand current patents and identify patentable opportunity
SHAPE	Identify how these features can be achieved at a reasonable cost and timescale to secure the commercial opportunity	Developed initial concepts, cost and technical review result in promising opportunities	Lab scale test results, patents
SCOPE	Internal and external resources working to deliver the opportunity	Cost and technical review, mechanical configuration, constructions and operations configuration	Pilot/demonstration scale results

Commercial status

	Status	Characteristics	Evidence
SHELL	Identified potential partners who are willing to explore a common theme	Concept development, relationship building	Defined program of work to gain entry ticket, successful PQQ partner application
SHAPE	Working with partners toward a common goal	Creating value, supporting a tender bid with significant resources	A joint team working on a response to a specific market opportunity
SCOPE	Working with partners toward an agreed deadline	Jointly delivering the program	Supporting a customer project

FIGURE 8.2
Managing technology/commercial lag.

8.5.4 Project and Risk Management

This is the stage of the process where there is an explosion of involvement. Make sure that the project team accountabilities encompass the delivery of business benefits and not just the physical or technical aspects of the product/service.

Project governance should include a measurement of how well lessons learned are transferred as vendors and project team members leave the project and hand over to operations.

Risk assessment at this stage should focus on risks to flawless operation from day one and potential supply issues for best-case/worst-case demand scenarios

8.6 IMPLEMENTING EPM

Start by assessing your organization's EPM capabilities.

Get the right people in the room and assess your most recent NPD project performance through the following three EPM steps using a 1–5 scale (1 = unstructured process, depends on individual insight; 3 = formal stable system in place and working, achieving flawless execution most times; 5 = significant history of success in reducing NPD time and raising NPD ROI).

- *Shell: Innovation*: How good is your organization at generating innovative product and service concepts that create new value and advantage for customers?
- *Shape: Idea development*: How well does your organization select winning ideas and develop them into practical product and service offerings?
- *Scope: Product/service design*: How successful is your organization at finding and working with strategic partners to translate development ideas into products and services that are easy to make and supply?

Involve representatives from commercial/customer-facing operations, and technology functions. Don't award yourself more than a 3 unless you have formal, measurable processes in place and working. A score of 4 is for those who can demonstrate at least one to two years' measurable improvement in ROI for at least two NPD areas.

Once you have assessed areas of strength and weakness, use a pilot project to address gaps in

- Design and performance management: The process of comparing options with standards to guide design decisions, tease out latent weaknesses, and develop the organization's capability to deliver ROI goals
- Specification LCC management: The process of selecting and refining winning ideas and translating them into low-LCC, high-value product and service offerings
- Project and risk management: The process of planning, organizing, and controlling resources at each stage to deliver flawless operation from day one and ROI
- Project governance: The process of allocating resources to winning product and service development

8.7 CHAPTER SUMMARY

This chapter cover the following points:

- What is EPM and why do we need it?
- EPM steps.
    - *Shell*: The innovation generator
    - *Shape*: Idea development
    - *Scope*: Product/service design
- Design, specification, and project management approaches to support each step.
- Links with EEM including EPM project governance issues.
- Implementing EPM.

The key learning points are as follows:

- The product defines the capital, which defines the operation.
- Integrate shape and EEM concept activities to improve the design for manufacture.
- Collaborate with strategic partners to refine good ideas into winning products and services.
- Balance commercial and technological progression through the steps to allocate scarce resources to product and service offerings with the best chance of success.

Appendix

EEM Design goals: Assessment criteria for 1 to 5 scale

	Goal	3. Acceptable	5. Optimum
 Customer quality	Process is able to meet current and likely future customer QCD features and demand variability. Provides a platform for incremental product improvement.	Easy order cycle completion Maximum control of basic and performance product features Flexible to product range needs	Capacity for future demand Robust supply chain Simple logistics/forecasting needs Flexible to potential market shifts
Life cycle cost	Process has clearly defined cost and value drivers to support life cycle cost reduction, enhance project value and maximize return on capital invested.	Clarity of current capital and operational cost drivers and process added value features Potential for value engineering gain Resource economy	High level of resource recycling Flexible to financial risks (e.g., vendor) Easily scalable to 400% or to 25% Access to high added value markets
 Operability	Process is easy to start up, change over, and sustain "normal conditions." Rapid close-down, cleaning, and routine asset care task completion.	Simple set up and adjustment mechanisms Quick replace tools Simple process control Auto load and feeder to fed processing	One touch operation for height, position, number, color, etc. Flexible to volume risk Flexible to labor skill levels
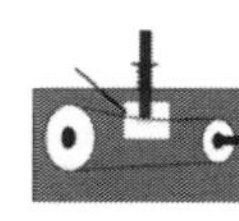 Maintainability	Deterioration is easily measured and corrected. Routine maintenance tasks are easy to perform and carried out by internal personnel.	Easy failure/detection/repair Off the shelf/common spares used Long MTBF, Short MTTR Easy to inspect and repair	Easily overhauled Self correcting/auto adjust Inbuilt problem diagnostic Predictable component life Fit and forget components
 Reliability	Function is immune to deterioration, requiring little or no intervention to secure consistent quality.	Low failure rate Low idling and minor stops Low quality defect rate Flexible to technology risks Good static and dynamic precision	High MTBI Stable machine cycle time Easy to measure Flexible to material variability
 Safety and environmental	Function is intrinsically safe, low risk, fail-safe operation able to easily meet future statutory and environmental limits.	Little non standard work Moving parts guarded, few projections Meets SHE and fire regulations Easy escape routes and good ergonomics	Foolproof/failsafe operation High level of resource recycling Uses sustainable resources

Index

F

G

H

I

Q

R

V

W